The Whole Earth Energy Crisis: Our Dwindling Sources of Energy

About the Book

There seems to be no limit to our expanding population's demands for energy. But there are limits to our most popular sources of energy. The time must surely come when our supplies of gas and oil, for example, will run dry. What will people do then? In this book, author John H. Woodburn opens up this disquieting question by telling the story of where the world's energy comes from and of the difficult decisions we face if the earth's resources are allowed to dwindle away.

This is the whole earth as seen from 22,300 miles in space. Among the troubles of the peoples inhabiting it is a crisis in the amount of energy necessary to comfortable, healthful life. (Courtesy of NASA)

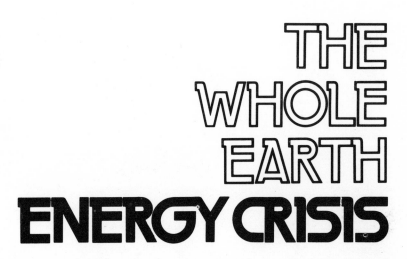

THE WHOLE EARTH ENERGY CRISIS

Our Dwindling Sources of Energy

by John H. Woodburn

G. P. PUTNAM'S SONS

NEW YORK

SBN: GB-399-60855-9
SBN: TR-399-20366-4

Library of Congress Catalog Card Number: 73-77422

PRINTED IN THE UNITED STATES OF AMERICA

12 up

CONTENTS

The Whole Earth Energy Crisis: Our Dwindling Sources of Energy

Everything that mankind does on earth requires energy. Here at the Kennedy Space Center in Florida is a dramatic example: the 900,000-gallon liquid-oxygen storage tank which services the launch vehicle for spacecraft. (Courtesy of NASA)

1 Energy Troubles Loom

The town clock in the public square of a picturesque Bavarian town is striking the twelve notes of noon. Townspeople and tourists have paused to look up at the ornate bell tower of the city hall. They know that carved and painted figures will appear soon and perform their noontime show. Musicians will make music. Cobblers will cobble shoes. Tailors will sew seams. Bakers will bake bread. Cabinetmakers will wield hammers and saws. Goosegirls and goatherds will tend their flocks. Kings and princesses, mayors and councilmen will appear.

When the clockwork machinery runs down, the show stops. This is not surprising. Everyone knows that windup toys will run only until they run down. The pretty scene on the bell tower happens only because someone wound up the mechanism, and the wheels turn and the levers move only until the stored energy is spent.

Actually what the real people are doing in the town square is much like the action of a gigantic windup toy. The same applies to every person, gadget, or animal that moves or causes something else to move throughout the whole earth. For energy runs the whole world of people, animals, plants, machines, gadgets—even the earth's clouds, winds, and waves.

But there's trouble ahead. The whole world of man's activities is using up energy faster than it is being replaced. Especially in the more highly mechanized countries of the world, the whole world of our activities is running down faster than it is being wound up.

Suppose the day comes when there simply isn't enough energy on hand to keep our whole earth toy running the way we have become accustomed to its running. What then? Will we live it up, have one last glorious fling in this affluent society that so many people already feel is a fool's paradise? Probably not. That is not what we expect from intelligent, socially sensitive people, especially young people who are already thinking about what they expect from and want to do in the world's giant windup toy of the future.

Few people seem aware of the threat of an energy shortage. Only rarely do we hear of brownouts, which happen when the available electric power systems are strained beyond their capacity. Only recently have gas companies had to tell potential new customers that they can't handle additional demands on their supply. Oil companies are still urging us to buy more and more of their products. So are coal companies. The huge bursts of energy released by atomic bombs worry us more because they might blow up the whole earth toy than because they point toward dwindling energy resources.

The very nature of energy makes it easy to take for

granted. It is invisible in nearly all its forms. An energy-loaded lump of coal looks no different from any other shiny black rock or mineral. A dead storage battery or dry cell looks exactly like the one that is ready to start your automobile or bring music from your pocket radio. A gallon of gasoline can look like a gallon of water. Most amazing of all, the energy in our food, which keeps our bodies warm, enables our muscles to work, fuels all the other energy-consuming needs of staying alive, is invisible and "unfeelable."

Many people aren't worried about the threat of an energy shortage because they remember from their school science courses an oft-repeated phrase that went something like "energy can be neither created nor destroyed." This suggests that the energy which keeps the whole earth windup toy running is somehow endlessly recycled, that the energy we are using now can somehow be recovered and used year after year.

But the simplified "conservation of energy" law is not the whole story. Even though energy may not be destroyed, it can get away. In fact, every time energy changes from one form to another, some fraction of the energy is lost forever in the sense that we cannot use it again. This is why last year's energy can't be recycled and made to do next year's work. This is why human and natural activity is using up energy faster than it is being replaced—why the whole earth toy is running down faster than it is being wound up.

The idea that the world's energy is on a one-way street is not easy to understand. Even advanced physics students have trouble grasping this idea. But the earth's energy resources, what and where they are, and how rapidly they are running out is everybody's business.

Actually, the threat of an energy shortage should make everyone want to have a better understanding of the nature of energy. There are responsible "energy experts" who provide leadership. One is the chairman of the U.S. Federal Power Commission, John N. Nassikas, who says, "Our resources are finite. Man's wants are infinite." What better way is there to let people know that the threat of an energy shortage is based on cold logic?

Nassikas goes on to say that "energy is the lifeblood of our economy and the catalyst enabling us to convert raw resources into the products and services that constitute our high standard of living." To back up his statement, he quotes from a study prepared to help members of the U.S. Congress meet their responsibilities as the nation's lawmakers. From this study Nassikas quotes, "The economy of the United States and the technologically advanced nations is based on energy. Energy is the ultimate raw material which permits the continued recycle of resources into most of man's requirements for food, clothing, and shelter. The productivity (and consumption) of society is directly related to the per capita energy available."

Having emphasized the role of energy in society, Nassikas goes on to say that the best way to head off the serious consequences of an energy shortage is to have the government provide regularity restraints to ensure that our energy resources are handled wisely. Here Nassikas

12

brushes up against the traditional idea that the earth's energy resources are available for unregulated, unrestrained development. He brings into the open one of the very complex aspects of the nation's energy problem —the question of individual freedom versus government control; the freedom of individuals to use their initiative (and good luck) to find, extract, and sell the earth's energy resources versus governmental control of the resources.

Arguments about individual freedom versus governmental control get tangled up with all kinds of social and political problems. Many such problems touch on beliefs and traditions that are set deep in a nation's history. In fact, conflicts between nations often hinge on the differences between individual freedom and government control.

These differences stand out clearly when we contrast the United States with the Union of Socialist Soviet Republics. Traditionally in the United States, people are free to turn their initiative loose and develop any resource or manufacture and sell any legal product. Not so in the USSR. This was brought out very clearly in an explanation of why Russian children try to barter with tourists for chewing gum. As a Russian tour guide said, "We don't have chewing gum in Russia, but our government right now is deciding whether or not to have chewing-gum factories."

Whether a nation lets its people have chewing gum may seem to have little to do with how a government responds to the threat of an energy shortage. It is more a question of how people respond to what their government does.

How do people react to the threat of an energy shortage? Some look favorably upon the dwindling supply of coal, oil, gas, and uranium. They think running out of energy resources will solve all society's problems. They seem to long for the good old days when mankind relied on muscles or those of his horses, oxen, or other beasts of burden. To these people, the machine age and its unlimited use of energy has brought with it such social problems as poverty, crime, boredom, erosion of civil rights, and pollution of man's habitat, especially the natural environment.

It is easy to see how this attitude has developed. It is too easy to get caught in a traffic jam or find the air almost unfit to breathe. Too much of the countryside is badly littered with the junk and trash of today and yesterday. Too many streams and beaches are ugly messes rather than pleasant parts of the landscape. The argument seems to be that when we no longer have abundant energy resources, these problems will disappear.

In the words of Nassikas, "This is a viewpoint which certainly deserves thoughtful study, but it should be analyzed in conjunction with a basic study of energy's contribution to our national goals for economic growth and improved living standards." This suggests that we must pay attention to social problems but at the same time keep human progress in mind. Nothing is to be gained by repairing a part of the whole earth toy at the expense of other equally worthwhile parts.

14

Especially while he was chairman of the U.S. Atomic Energy Commission, Glenn T. Seaborg gave much thought to the threat of an energy shortage. He deals with the above argument when he says, "There are some who see the cooperative use of vast amounts of energy— energy supplying unlimited power across national borders, raising living standards, lowering production costs, helping to distribute water and produce food—who see all this as a prime source of achieving a new era of peace and progress in the world. There are others who see abundant energy as having only a marginal effect in promoting human progress. And there are still others who see the future use of more energy as devastating to man and the whole planet."

As Dr. Seaborg sees things, "To put it quite frankly, many people today—because of the focus on environment—have turned on energy as a major villain of our time. Whereas only a few years ago its laborsaving, life-saving and mind-stimulating attributes were praised, today energy, and particularly the idea of its continued growth, are viewed as threats. Our ability to release and apply massive amounts of energy is seen as a first cause in the ultimate destruction of the earth. Whether it is the potential destruction of nuclear weapons, the air pollution of combustion, the thermal effects of power generation, the radioactive effluents of nuclear power plants, the waste produced by industry, the indiscriminate scraping of a bulldozer or the noise of a jet, many people are pointing the finger at energy, or more precisely our command and release of it, as the real culprit. I think

most of us know that to make energy as such the scape-
goat for man's indiscretions, excesses, or just stupidity, is
quite naïve. It is much the same as saying that if we did
not have food we would not have indigestion."

Speaking more as a concerned citizen than a government official, Dr. Seaborg goes on to say, "In view of both
the longer-term energy resource problem we face and the
shorter-term environmental crisis due to the misuse of
such resources, it seems to me that the time has come—or
perhaps is overdue—when men, nations and the entire
international community must begin more serious energy policy planning. If we are not to thoroughly deplete
many of our irreplaceable natural resources in a matter
of a few generations, and at the same time put those
generations in environmental jeopardy by the misuse of
those resources, we must begin to use those resources in a
highly rational way."

There will be people who will resist the very idea of
being told they can or can't have as much energy as they
want for however they want to use it. To be told that we
will have to cut down our use of energy would be a real
shock. For years, especially in the United States, people
have been encouraged to use more and more energy.
Government policies have backed up the sales efforts of
the energy industry to make abundant supplies of energy
available at the lowest possible prices.

S. David Freeman, writing in the *Bulletin of the
Atomic Scientists* for October, 1971, calls this "a promo-
tional era in energy growth." But now that we have become accustomed to our electric toothbrushes, air-condi-

tioned homes, high-powered automobiles, and all manner of energy-hungry luxuries and necessities, Freeman believes that "the alarm bells have begun to ring and we fail to listen to them at our peril." He calls for an era of energy conservation. The shortage of energy resources and the increasing seriousness of the environmental pollution problem call for everyone to become energy-conservation conscious. To Freeman, turning off the lights when we leave a room or turning down the thermostat when we go to bed isn't something to joke about. He is convinced that we all could live just as well as we do now and yet use much less energy if we learned to use it wisely.

Freeman also says, "The automobile really highlights our waste of energy. We power a two-ton vehicle which is less than 10 percent efficient in the use of energy to transport a single person. We need more electric cars, large ones on tracks, to conserve the massive resources now being wasted in the transportation that is jamming our cities' streets and polluting the air more than any other source."

There are good reasons for putting the impending energy shortage and the problems of environmental pollution in the same package. They are related. Unfortunately, neither problem gets any simpler when it becomes entangled with the other. Nor is it any less urgent that we get on with finding solutions.

Michael McCloskey, executive director of the Sierra Club, is against anything that threatens to damage or destroy our natural environment. And he doesn't pull his punches when dealing with the energy problem. He is

17

very much against advertising campaigns and government policies that urge us to use more energy. To McCloskey, a nation shouldn't strive to increase its gross national product every year simply for the sake of growth itself.

According to McCloskey, ever-increasing growth rates run into all sorts of limits. He uses the rate at which new power plants are being built to prove his point. The number of power plants needed to provide electricity in the United States doubles every ten years. A typical power plant occupies an area approximately 1,000 feet on a side. If we were to double every ten years the amount of land devoted to power plants, McCloskey says that "in less than twenty doublings—less than 200 years —*all* the available land space in the United States would be occupied by such plants."

Obviously no society would ever permit its landscape to be taken over by power plants. Calculations such as these, however, are often used to emphasize the urgency of social problems.

Continuing his action program, McCloskey believes that we will have to ban advertising, government subsidies that lower the apparent price of energy, and all other efforts to get us to use more energy. In addition, he asks for legislation and regulatory action that will head off oil spills, do away with strip-mining, put stricter controls on the disposal of nuclear wastes, and cut down the number of automobiles, power plants, and industries that consume fuel.

McCloskey wants us to be more considerate of future

18

generations and set aside some of our dwindling energy resources, as well as increase our efforts to repair the damage of our undisciplined use of energy to their air, water, soil, and landscape. We are to look upon high-powered, energy-hungry automobiles and gadgets that needlessly replace muscles with electric power as symbols of the squandering of our earth's energy resources. Proper status symbols in the future would be riding bicycles, turning off the lights when we leave a room, closing doors and windows when furnaces are on, and using elbow grease rather than kilowatts whenever possible.

Action programs that attempt to establish the "in thing" to do can be very effective albeit difficult to launch. But in the future we may see programs encouraging us not to replace useless gadgets.

Another point in McCloskey's action program calls for science and technology to come up with more efficient ways to use the energy we must use. Today's automobiles, furnaces, air conditioners, power plants—in fact, nearly all of today's energy-hungry gadgets—are embarrassingly inefficient. Much of the energy in fuels goes up smokestacks or out exhaust pipes. Much electrical power becomes unwanted heat. In theory, the efficiency of our energy-consuming gadgets could be doubled or even tripled. This would not only stretch our energy resources but also cut down the harmful effects of energy consumption on the environment.

To McCloskey, some industries are worse offenders than others when it comes to using more energy than they should. The planned obsolescence of autos, for example, the campaigns to trade in last year's model for a newer albeit no more efficient this year's model makes little sense from an energy-conservation point of view. He also criticizes the aluminum, paper, and fertilizer industries for being slow to promote recycling or replacing energy-hungry items with those that use less energy to produce but do their jobs just as well.

Finally, McCloskey calls for greater efforts on the part of the government to control and oversee the production, distribution, and use of energy and to tie the use of energy resources with responsibility for maintaining the environment.

Wars and the Whole Earth Windup Toy

Each shell that is fired, each grenade thrown, each bomb dropped spends energy. There can be no recycling of the energy used to destroy people and materials. Factories, steel mills, mines and quarries—indeed, the nation's whole industrial and distributive machinery—become increasingly energy-hungry in time of war. Waging war is a big part of the world's gigantic windup toy. But it isn't easy to study the link between the impending energy shortage and the hope that someday nations will no longer resort to war to settle conflicts.

We could say that one way to stop war is to let the whole earth's energy supply dwindle to a point that would make war impossible. But then we realize that wars have been waged even when energy demands were very low. We could also look upon the proper manage-

20

ment of the world's energy resources as the challenge we need to get the world's peoples to live together in peace.

Energy resources will certainly play increasingly crucial roles in international relations. Discussions of past wars often feature the idea of "have" versus "have not" nations. The future may find us with "have had" as well as "want to have" nations. It's true that some nations are using up their natural energy resources much faster than others and that the world's energy resources are not distributed equally. Some nations control far greater amounts of available energy than others. Furthermore, some nations have been slow to develop their energy resources; others have moved full speed ahead and are well along toward having burned up their supply of gas, oil, coal, or uranium.

Nations slow to develop their energy resources are sometimes called the underdeveloped nations. This suggests that overall standards of living in these nations are lower than in overdeveloped nations (inasmuch as a nation's standard of living seems to be linked with its per capita energy use). This can lead to very sticky international questions. For example, will nations that have put up with low standards of living be willing to share their not yet developed energy resources with those nations that have enjoyed high living standards and are consequently running out of energy? Could these underdeveloped nations decide that it is now their turn to become affluent? Or could the lack of cooperation between the "have had" and "want to have" nations increase the threat of world war?

The Whole Earth Toy Is Growing

Not only are people asking for more energy each year, but each year finds us with more people. Not only are we running out of what it takes to wind up the whole earth toy, but each year finds the toy becoming more complicated, with more bits and pieces calling for their share of energy. On this basis, plans for the intelligent use and thoughtful conservation of the world's energy resources must anticipate how many people there will be on earth by the time the plans are to be carried out.

Attempts to estimate population trends are based in part on how rapidly people are reproducing today and how rapidly previous generations reproduced. Let's play a bit with the problems of population estimates. In the United States today the average life-span is about seventy years. To make the arithmetic easy, let's estimate the total population to be 210,000,000 and assume that there is the same number of people in each age bracket. On this basis, 3,000,000 people can be expected to die each year.

If the total population in the United States were to stay at 210,000,000, there would need to be 3,000,000 births each year. There would be, supposedly, 105,000,-000 couples in the total population. Assume that only those couples between the ages of eighteen and thirty-eight make up the childbearing age-group. This age-group represents three-tenths of the total population: that is, 31,500,000 couples. For the total population to

22

stay at 210,000,000, fewer than one-tenth of these couples could have children in any one year. Furthermore, to hold the total population constant (assuming that all couples were to have children sooner or later), no couple could have more than two children.

Such birthrates are not the usual state of affairs. Actually most human populations are increasing and the problem is to estimate the rates of increase. In attacking this problem, scholars have tried to guess how many people have lived since the origin of mankind. One estimator says there were 1,000 people in 1,000,000 B.C. By 10,000 B.C. he estimates the total world population to have been 1,000,000, and by 0 B.C., 275 million. By 1970, the world population had reached more than 3 billion people.

These data suggest that in the very early history of man, it took 100,000 years for the world population to double. Near the advent of the Christian Era, world population doubled every thousand years, and it has been doubling in a smaller number of years ever since. Using world populations as estimated by the United Nations for the years between 1950 and 1960, world population doubles nowadays in approximately thirty-eight years.

If nothing happens to slow this rate of increase, by the time a child has lived scarcely more than one-half a lifetime, the world would be twice as crowded as when the child was born and four times that crowded by the time of old age.

In the language of a growing energy shortage, within a lifetime, a person would find four times as many people pulling their cars up to gasoline filling stations, plugging their electrical gadgets into outlets, pumping oil or gas into their furnaces, and doing all the other energy-consuming things we take so much for granted.

Recall the two short sentences that so crisply described the origin of the energy crunch: Our resources are finite. Our wants are infinite. In a sense our ability to increase the population is also infinite. In fact, this may be why energy needs are infinite. If this is so, when we face the problems related to maintaining adequate supplies of energy, we will also have to tackle the equally sticky problems associated with worldwide population growth.

Can the Whole Earth Toy Slow Down?

Suppose we have to cut down the amount of energy each of us is permitted to consume. What would it mean? How would we have to change the way we live, the way we are accustomed to doing things? Would today's people be able to adjust to using no more energy than people did a hundred years ago? Or five hundred? Or a thousand?

Roll back the calendar a hundred years. At that time, an average person in the United States consumed about 5 gallons of petroleum per year; today's average annual per capita consumption is more than 1,000 gallons. Today's person uses three times as much coal as in 1870; and the average person then scarcely knew what natural gas was. All in all, the average person in the United States today consumes between seventeen and eighteen

24

times as much energy as he did in 1870. Although this is a much greater rate of increase than is the case worldwide, it is true that the whole earth toy consumes many times as much energy as it did in the not too distant past.

Roll back the calendar a hundred years and look at man's accomplishments and the way people lived. Look at cathedrals and castles, paintings and sculptures, highways and bridges, theaters and sports arenas, printing presses and wineries, spinning mills and foundries. All this was done before gas and oil had been much more than discovered and the earth's coal supplies much more than sampled. Although coal was introduced as an energy source as early as 1500, wood provided well over half the energy needs throughout the world as late as 1870.

The discovery of the seemingly unlimited energy from coal-burning had much to do with sending man headlong into the Industrial Revolution. But there is a kind of satisfaction in not losing sight of human accomplishments before the machine age. It is reassuring to realize that even without the massive injections of energy that have come from burning the earth's store of coal, oil, and gas, man has been able to produce art and architecture, music and factories, bridges and boats. It is comforting to realize that man has always been able to find within nature's resources the wherewithal to keep fed and clothed, housed and enlightened.

Of course, this kind of comfort and reassurance is painfully inadequate for many prickly problems in today's society. It is hard to believe that the stream of civilization can ever slow down. Serious troubles would surely develop in any society were its people forced to cut back their energy consumption. There is little doubt that the withdrawal symptoms accompanying a shortage of work-saving, timesaving, standard-of-living improvement gadgets and products could throw a whole society into chaos or revolution.

People Aren't Equally Energy-Hungry

The peoples of the world get along with widely varying amounts of energy. In India, for example, people use only one-fiftieth as much energy each year as we do in the United States. People in England use twice as much energy as the people of the USSR. People in Japan use three times as much energy as people in Africa.

Obviously people can live even today without consuming energy at the rate we do in the United States—or in England, or Russia, or Japan. This fact provides another kind of reassurance against the threat of severely dwindling supplies of available energy. But this may not satisfy many people. Those who live where the per capita use of energy is low simply don't live the way people do in high-energy-consumption countries.

It isn't always easy to sort out the effects of differing rates of energy consumption on the way people live. To compare Japan's Expo-70 display of technological and cultural achievements with those of Canada's Expo-67, for example, doesn't necessarily reflect the threefold difference in their energy-consumption rates. In both cases there were crowded trains, bright lights, and the hustle and bustle of busy people. In both cases the people were well clothed, well fed, and apparently equally healthy.

But there were many little differences. At lunchtime, the rice balls eaten by the Japanese families simply seemed to appear and then disappear. In contrast, a typical American family's lunch came in all kinds of wrappings and containers, and there was as much evidence of their having had lunch when it was over as before they sat down to eat. All this evidence involved energy consumption. Many Japanese people carried moistened washcloths to wipe away the sweat and soil from the warm, dusty Exposition setting. In contrast, Americans were more likely to use throwaway paper tissues or make frequent trips to lavatories and use much running water, soap, and, again, throwaway paper towels.

These are seemingly trivial things. But the point is that in both Japan and Canada people solved their everyday problems of living—and apparently equally well. But their solutions revealed marked differences in energy consumption. When the crisis of worldwide energy resources comes, we may have to pay attention to these differences in the way people live.

One can see tall, modern high-rise buildings being built in Moscow, Washington, London, or Bangkok. At first glance each building seems to be the creation of the same energy-hungry bulldozers, cranes, and other massive construction equipment. A closer look brings out differences. In Washington each bundle of bricks, each window is hoisted to where it is needed by powerful machines. Sometimes the same giant crane is used to lift a many-ton steel beam or a handful of rivets. In Moscow, however, one sees brawny men and women using pulleys to hoist, hand over hand, the bricks or mortar they need to finish a building after the heavy equipment has moved on to new construction sites. And in Bangkok, although the façade of a new high-rise office building looks very much like similar buildings in Washington, wiry little women use hoes and shovels to mix the mortar they need to put the finishing touches on the tile and marblework inside.

High-rise buildings, rice balls, hot dogs, facial tissues, massive construction equipment, laundered washcloths, hand-over-hand pulleys, and giant cement mixers seem a strange jumble of words to discuss an impending energy shortage. In fact, the jumble could include thousands more words—the thousands of words it takes to describe the gigantic windup toy that is the whole of human activity—the worldwide array of activities that are powered by the whole earth's energy. For it is energy that ties together all activity. The earth's energy resources know no political boundaries. Man's dependence on energy doesn't change when we stop to have passports checked or to shift from one foreign-language phrasebook to another. Nor is man's dependence on energy a sometime, one-time thing. The whole stream of civilization has always been powered with energy. It always will be.

The story of civilization is the story of man's use or abuse of the earth's energy resources—the happy story of

28

wise use and conservation of the earth's energy or the tragic tale of unwise, selfish waste. The few thousand years of the development of civilization is only a very, very small fraction of the total expanse of time. There is a long, long way to go. We would like to believe that the accomplishments of mankind will continue to rise to ever and ever greater heights.

This can't be done with finite energy resources or without supremely enlightened planning. The way things are going today, the world's fossil fuels—coal, oil, and gas—will have played out their roles in the whole earth windup toy in only a fraction of the total expanse of man's existence. These fossil fuels have been 500,000,000 years in the making, but they could all be burned up in 300 years. What then?

2 Energy from the Sun

The story of the whole earth's energy begins with the sun. And this may well be where it will end.

The sun's heat and light were the only known sources of energy until only a few hundred years ago. Even today the sun provides the biggest share of the energy that keeps the earth and everything on it going. This is true despite the thunder of jet engines burning tons of fuel, or the earthshaking shock of rockets blasting off into space. Or the constant din of auto and truck engines burning thousands of barrels of gasoline and diesel fuel. Or hundreds of steam-generating furnaces swallowing trainloads of coal.

Nearly 100 times more energy is spent driving nature's solar engines than is consumed by all fossil-fuel-burning engines. Only a fraction of the sun's energy is spent in driving the earth's air masses; in forming and moving its clouds, its ocean waves; in driving the green plant world's photosynthesis reactions; and in warming the earth's atmosphere, its waters, rocks and soil. It would take 445,000 times as many electric generators as there

31

are in the United States to produce as much energy as the sun.

The Magnitude of the Sun as an Energy Source

At the outer limits of the atmosphere the sun's energy streams toward the earth at the rate of approximately two-tenths calorie per second on each square inch of surface. If there were some way to change solar energy to food energy with 100 percent efficiency, a person could live on the sun's energy that falls on about five square feet of surface at the outer limits of the atmosphere. The "100 percent efficiency" is a big if. In fact, it is less likely than a person surviving at the outer limits of the atmosphere. Today the only way to feed solar energy into the metabolism of our life processes is by way of the fruits, grains, and vegetables we eat or feed to meat-producing animals.

When we consider the efficiency of the processes that provide our food, we find that a person consumes the energy salvaged from the total solar energy that falls on 40,000 square feet of the earth's surface—an area slightly smaller than a football field. This estimate is based on an average of the sun's energy that falls on the whole earth's surface, night and day, year in and year out. If we think of the sunlight falling at high noon in a desert climate on a summer day, a much smaller area would provide a person's 3,000,000 daily food calories, or, as more commonly stated, 3,000 kilocalories.

The sun is so enormous and produces such a fantastic

amount of energy that it stretches our imagination to understand solar information. For example, the earth receives each day nearly 200 times as much energy from the sun as is consumed during a full year by all the energy-using activities in the United States. In other words, if we think of everything that goes on in the United States as being a windup toy, the earth receives enough energy from the sun in one day to keep 200 such toys running for a whole year.

Each year the sun sends toward the earth 30,000 times as much energy as is used by the entire world's industries. Or 5,000 times as much energy as is stored in all the earth's volcanoes, hot springs, and similar geothermal phenomena. Or 60,000 times as much energy as is spent keeping the oceans' tides ebbing and flowing.

M. King Hubbert, writing in *Scientific American* for September, 1971 reports the sun's total energy falling on the earth to be 1.73×10^{17} watts. On this basis, if all of the sun's energy received on the earth in one hour were converted to electrical energy and run through a household electric meter, the meter would read 173 trillion kilowatt hours.

Hubbert also helps us understand what happens to this energy. For example, slightly less than one-third of this energy is merely bounced back into space. These 50 trillion kilowatt-hours are lost insofar as our whole earth toy is concerned—with the exception of giving astronauts a magnificent view of the earth.

Much of the sun's energy input goes to warm the earth. Once this energy is changed into heat, however, it is radiated back to space. If an imaginery electric meter were used to keep track of the earth's heat bill, the meter would read 80 trillion kilowatt-hours. For comparison, if the world population were one billion families, and if each of these families used electricity at the rate of typical U.S. suburban families, the billion meters would add to only a little more than 50 trillion kilowatt-hours.

Four trillion kilowatts of the sun's energy is spent keeping the earth's waters moving in cycles from clouds to precipitation and back to clouds. If we ran the energy spent driving the winds, waves, and ocean currents through the same meter, another 370 billion kilowatt-hours would be added.

If the world of green plants were to pay a solar energy bill, and all the energy that drives photosynthesis were to go through an imaginary electric meter, the meter would read 40 billion kilowatt-hours. This is the fraction of the sun's energy that affects our lives most directly.

At the Source of the Sun's Energy

It is interesting to try to trace the sun's energy to its source and to imagine what kind of fantastic furnace it is that puts out this enormous amount of energy. Start at the outer limits of the atmosphere. Here on earth enough solar energy falls on an area the size of a postage stamp to heat a spoonful of water about two Fahrenheit degrees in one minute. Track this heat back the millions of miles to the surface of the sun. There the same postage-stamp-size area would receive over 200,000 times as many calories each minute. This means that a spoonful of water would not only be boiled away instantly, but it would also be separated into its elements. And so would the spoon!

Estimates of the conditions that exist at the surface of the sun are based on clear-cut information and logic. The inverse-square law is a good example: As light spreads out from a source, its intensity decreases as the square of the distance increases. If you move twice as far from a light source as you are now, the amount of light falling on you is only one-fourth as great as when you were twice closer.

The distance between the sun and earth is known. So is the intensity of the energy that reaches the earth from the sun. With these things known, it is possible to use the inverse square law to calculate the "unknown" intensity of the sun's energy at the source.

There is another way to estimate the temperature of the sun at its surface. This method is based on the fact that hot objects send out energy at a rate proportional to the fourth power of their absolute temperature. We assume that this relationship holds for the sun in space. It is known that the sun sends out slightly less than 10 calories of energy per second from each square inch of its surface; it follows that the temperature of the sun's surface is approximately 10,000° F.

Knowing both the intensity of the energy radiated and the total surface area of the sun, it is possible to estimate the amount of energy given off by the sun: It is 88 septillion calories per second, an incomprehensibly large

35

amount of heat. Even so, the human body is a better furnace than the sun. This puzzling statement is based on comparing, pound for pound, the heat-producing actions of the two bodies. The sun has such an enormous mass (two decillion grams, or more than four quintillion pounds) that its rate of heat production is approximately two-thousandths of a calorie for each hundred pounds. A person's body produces heat at a much greater rate.

This isn't so puzzling when we realize two things. First, the volume of a person's body is almost infinitely smaller than the volume of the sun, and second, there is relatively more surface on a small object than on a large one. For example, if the radius of one sphere is twice that of another, the larger sphere has eight times as much volume but only four times the area of the smaller. To determine the volume of a sphere, the radius is cubed. To determine the area, the radius is squared.

These ideas produce interesting results when we compare a sparrow with an elephant. Pound for pound of body weight, the elephant gives off only one-thirtieth as much heat as the sparrow. If the elephant produced heat at the same rate as the sparrow, its greater mass would build up heat until we would have roasted elephant.

The Source of the Sun's Heat

Modern theories say that the sun's heat is produced by nuclear events—by a kind of "hydrogen bomb" that explodes continuously near the center of the sun. At the center of the sun, although the process occurs in more than one step, four hydrogen atoms are brought to-gether, or fused, to make one helium atom. When this happens, mass is changed into energy. The "hydrogen" or fusion-type atomic bomb takes advantage of this reaction.

When H. A. Bethe put together his theory to explain the sun's energy, he created a classic example of the workings of a scientist's mind. From his knowledge of nuclear reactions in general, he developed a theory that agreed with everything that was known about the sun's makeup and the rate at which it gives off energy. Most important of all, Bethe had the creativity it takes to imagine how something is happening 93 million miles away.

If today's theories regarding the origin of the sun's energy are correct, and they seem to be, the sun will continue to send energy toward the earth as it does now for another 30 billion years. This is why we say the sun is a finite but inexhaustible energy source.

Solar Energy and the Plant World

In a very real sense, man is simply one of the many different kinds of living things in the whole earth ecological system. Although man plays around with the whole earth toy, adding bits and pieces here, tearing down bits and pieces there, speeding up or slowing down the action of one part or another, he doesn't really run the whole show. With few exceptions the activities of all living things depend on hitching rides on the flow of solar energy through the earth's environment. The exceptions are the activities that depend on tidal forces, geothermal energy from hot springs or volcanoes, or nuclear energy.

In a sense, living systems snag their share of the sun's energy as it goes by. This is the energy that keeps them alive or is stored for future activities or the activities of their offspring. Man has many ways to increase his share of the sun's energy, to divert the flow of the sun's energy away from its natural path, and to run it through the things he wants to do. Man is not content merely to be born, eat, grow, move about, reproduce, and die. Man is a builder, an artist, a creator of things and ideas. Man likes to go places and do things—all kinds of places and all kinds of things. But everything that man sets out to do demands its own energy quota—its own piggyback ride on the flow of energy through the environment.

Before man discovered the fossil fuels—coal, oil, and gas—activities were limited to those that could be carried on with the energy captured from each day's rising sun. Having discovered the energy that can be released by burning coal, oil, and gas—and more recently, by the fission or fusion of atomic nuclei—man went ahead to set up new flows of energy through the environment, in addition to the flow of solar energy. This means that we have become accustomed to using more energy each day than we are harvesting from the sun—building more buildings, creating more things and ideas, painting more pictures, sculpting more sculptures, composing more music, and doing countless other energy-demanding things.

38

What would it mean if the whole earth toy were to be wound up once again with only the daily input of the sun's energy? What would it mean if human activities were limited to those that could be powered by contemporary, not fossil, solar energy? Obviously, these questions deal with a state of affairs so utterly improbable that they are not meant to be answered. But they are meant to emphasize that man is very dependent on capturing the sun's energy as it flows through the environment. This is where the world of green plants plays a vital role. It is only through the photosynthesis processes of green plants that the sun's energy can be built into the food supplies for all living systems.

The 40 billion kilowatts of solar energy involved in worldwide green-plant photosynthesis arrives in the form of light waves. Each bit, or quantum, of energy has a definite set of characteristics. We are most familiar with the set of light waves that make up the visible spectrum —the mixture of colors of light so beautifully displayed in the rainbow. We are less familiar with the invisible waves that are neighbors of the visible spectrum, the infrared and ultraviolet spectrums. However, heat lamps and sun lamps are becoming common, and these lamps are designed to produce infrared and ultraviolet waves.

The sun's energy arrives at the earth mostly in the form of infrared, visible, and ultraviolet energy waves. By a series of transformations, it is ultimately reduced to thermal, or heat, energy. The sun's incoming energy can be converted 100 percent to long-wavelength, low-temperature heat energy. But this process is not reversible. Heat energy cannot be converted 100 percent to any other form of energy.

Furthermore, every energy transformation allows at least a fraction of the energy to escape in the form of heat. And everything that happens in the whole earth toy involves some kind of energy transformation. It is because energy is being changed from one form to another that the winds blow, the oceans' currents flow, tides and waves come and go, the rains fall; seeds germinate, plants grow, flower, bear seeds, and die; animals are born, grow, reproduce, eat, move about, and die.

To us, man's activities are the most important part of the earth's gigantic windup toy, the whole earth ecological system. Human life processes begin with the capture of solar energy and its storage in green plants. Being both a herbivore and a carnivore, man eats both plants and animals. But no food chain of meat-eating animals puts any animal very far from one or another herbivore. The whole world of plant and animal life (including, of course, mankind) is powered by a flow of energy that begins when green plants snag some of the sun's energy as it goes by.

In more formal language life depends on photosynthesis—that vital process carried on in green plants in which the energy of the sun builds carbon dioxide and water into energy-rich carbohydrates, with oxygen as a byproduct. The reverse of this energy flow sustains the entire system of plants, herbivores, carnivores, parasites—in other words, the whole earth living windup toy. Photosynthesis leads to energy-rich carbohydrates, proteins, fats, and countless other complex substances in the bodies of plants and animals. Oxidation, the reverse of photosynthesis, leads to energy-poor water, carbon dioxide, and other waste products.

In an average year green plants build more than 125 trillion pounds of carbon into new plant tissues. Seven times as much carbon is taken from the environment and used to make new plant tissues as is burned each year by the whole earth's use of fossil fuels. Although more than half these new plant tissues are produced by aquatic plants, more than 40 trillion pounds of carbon are built each year into the tissues of land plants.

On a bright sunny day a cornfield might receive 4,000,000 calories of solar energy on each square yard. Because photosynthesis is not a very efficient energy-transformation process, only 200,000 calories of the day's input of energy can be stored in the leaves, stalks, roots, and, eventually, the grains of ripened corn. In ten days a square yard of this cornfield would produce enough food to keep a person alive for one day. But who wants to live on a diet of nothing but corn? If you prefer beef, there are more problems. For example, a beef animal converts into meat only about one-tenth the energy in the corn it eats. This calls for a ten times larger cornfield.

To give a person 2,000 kilocalories of food energy calls for the energy in enough beef to provide 30,000 calories. The beef cattle will have eaten enough corn to give them 300,000 calories. And it took 30,000,000 calories of solar energy to grow the corn. It takes a whole acre of corn to catch this much of the sun's energy in one day.

Estimates of the efficiency of the photosynthesis process range from 0.2 to 3 percent. Under ideal conditions efficiency could be as high as 35 percent. Although men and women have always been interested in learning more about photosynthesis, the threat of dwindling supplies of other energy resources gives added reason for more research in this direction.

Interesting work along these lines has been done by a team of scientists composed of Edgar Lemon, D. W. Stewart, and R. W. Shawcroft. These men designed instruments and planned experiments that tracked the sun's energy falling on a typical growth of green plants, a cornfield. Their approach brought together the knowledge of botanists, meteorologists, physicists, and engineers because they knew that a variety of events and circumstances are involved in photosynthesis.

To deal with the complex mix of factors and forces driven by the sun's energy when it falls on green plants, this team of scientists designed a computerized model that duplicated as nearly as possible the conditions existing during photosynthesis. Because the model took into account the conditions of the soil, the plant, and the atmosphere, the scientific team named the model SPAM. SPAM turned out to be a very productive laboratory assistant. It recorded the amount of solar energy that fell on the plants in an experiment, the amount of this energy used to keep the plants alive, the amount stored in newly forming plant tissues, and the amount lost to the environment.

Dr. Lemon and his team reported many of their findings in the October 22, 1971, issue of *Science*. They found, for example, that on a summer day in the eastern United States, from 1 to 5 percent of the energy falling on a cornfield is caught up in the actual process of food manufacture inside the green leaves of corn plants. Between 40 and 50 percent of the sun's energy is spent in evaporating water from the surface of the leaves. From 10 to 60 percent goes to heat the air and soil.

SPAM was particularly useful in interpreting the results of experiments designed to change these percentages. It is known, for example, that water enters leaf tissues through tiny pores called stomates. Guard cells beside each stomate control the passage of water vapor, carbon dioxide, oxygen, and other gases between the moist tissues inside the leaf and the relatively dry air outside. By influencing the opening and closing of the stomates, one might decrease the amount of the sun's energy that does nothing more than evaporate water from leaf surfaces.

The action of the stomates is interesting for another reason. Supposedly, the more carbon dioxide that passes from the air through the stomates and into the inner tissues of the leaf, the more carbon will be made available for the synthesis of carbohydrates. SPAM made it possible to design experiments involving the amount of carbon dioxide in the air and the speed of the winds moving the air across the leaf surface. The computer involvement in the experiments made it easy to calculate the effects of having the stomates respond to various moisture conditions and at the same time to investigate the role of carbon dioxide intake.

The efficiency of photosynthesis is also affected by the

angle at which the leaves are exposed to the sun and the chances that one layer of leaves might be shaded by others. Supposedly, a farmer could increase the yield from a cornfield if he arranged the rows of corn so as to take advantage of the angle of the sun's rays. And a variety of corn that tips its leaves at one angle might be more efficient in certain situations than would another variety.

Dr. Lemon and his team found that attempts to improve the efficiency of photosynthesis by increasing the amount of carbon dioxide in the air are only partially successful. Relatively small increases of carbon dioxide in the atmosphere produce almost as much effect as tremendous increases. Supposedly, if we continue to burn fossil fuels at the current rate, the carbon dioxide added to the atmosphere could improve the efficiency of photosynthesis as much as 10 to 20 percent during the next hundred years.

Because it could handle a complex set of interdependent variables, SPAM was a great help in understanding how photosynthesis fits into the whole scheme of man's energy-consuming activities. Its data are useful in helping mankind plan what must be done to head off or adjust to the problems that are sure to come with an energy shortage, especially the all-important problem of adequate food production.

The Direct Capture of Solar Energy

There is a group of scientists and engineers who believe that there could be a more direct approach than relying on the world of green plants to catch and store a very small fraction of the sun's energy. To these scientists, the enormous amount of energy that comes to the earth from the sun holds the promise of solving the whole earth's energy-resource problems. Even though the sun's energy comes very much diluted—it's turned on each day and off each night and is easily interrupted on cloudy and stormy days—some scientists are convinced that solar energy can be collected and concentrated in convenient packages. And they believe this can be done economically in the near future.

When pessimists face the problem of collecting and concentrating the sun's energy, they think of the enormous array of mirrors or lenses that would be needed. They worry about the difficulties of storing heat when the sun isn't shining. When optimists confront these same problems, they think of the new discoveries taking place in the field of lightweight, cheap plastic substitutes for glass mirrors and lenses. They are encouraged by recent research on more efficient ways to store heat. Most of all, they know that the sun is an inexhaustible source of energy, and they feel that more and more people will join in their optimism when the shortage of other energy resources seems to be getting critical.

Norman C. Ford and Joseph W. Kane represent the optimistic approach. Their plans for harvesting the sun's energy begin with mass-produced plastic Fresnel lenses. In the October, 1971, *Bulletin of the Atomic Scientists* they explain how it would be possible and economically feasible to build a collecting system of plastic lenses more than a square mile in area. They have planned a solar-energy-collecting system that uses about two square miles of collecting surface. An area this large will collect enough solar energy to generate as much electricity as a modern electric-power station.

In the system planned by Ford and Kane, the Fresnel lenses focus the sun's energy on water in appropriately designed boilers. The water is heated to about 1500° C. At this temperature some of the water breaks down into its elements, hydrogen and oxygen. To separate the hydrogen from the hot steam, the boilers have windows covered with a membrane that has millions of tiny pores in each square inch of surface. These pores are too small to allow particles larger than hydrogen molecules to pass through. On this basis, the solar-energy-fired steam boilers produce hydrogen gas, an almost ideal fuel for various energy-converting processes.

Each square yard of the lens-covered collecting-surface is expected to yield about one-third pound of hydrogen per day. The cost would be greater than for an equivalent amount of present-day fuel. But the early stages of many kinds of industrial development are likely to be expensive. Furthermore, it is the threat of running out of present-day fuels that prompts solar-energy research. Ford and Kane believe that if hydrogen gas could be produced in large quantities, it would be a very good fuel: It is nonpolluting, and it can be liquefied—this suggests that it might be a source of fuel for automobiles and trucks or ships and airplanes.

Aden and Marjorie Meinel have also planned a system

for collecting solar energy and using it to replace other energy resources. They expect to trap the sun's energy by having it fall on a newly developed type of light catcher. They propose to spread out large areas of a material that consists of several layers of very thin films of molybdenum, aluminum oxide, or similar metallic substances. They know that when the sun's energy is reflected between these metallic films, the energy is absorbed and temperatures as high as 1000° F. are created. The Meinels plan to harvest the heat energy responsible for these high temperatures by having it melt substances—in much the same way that heat is stored in melted ice. These hot, melted substances would then be used to fire steam boilers or sent to other types of energy-converting processes.

To match the energy that is "for sale" at a typical billion-watt electric-power station, the Meinels' system would need a collecting-surface approximately seventy-four miles on a side. If we were to rely on such a method to collect and concentrate the sun's energy, thousands of square miles of the earth's surface would have to be set aside for the production of solar energy. This could be the price we will have to pay for the energy needed to keep the world of man's activities going on after the fossil fuels have been exhausted. But as the Meinels point out, this type of solar-energy plant would very probably be located in desert areas. And if an abundant source of cheap energy becomes available in these areas, perhaps it will be possible to desalt seawater or bring water into the area by some other means and thereby improve the environment enough to compensate for the space taken over for the collection of solar energy.

The challenge of using the sun's energy to replace other fuels arouses the inventive spirit, especially in countries where there are no natural deposits of coal, oil, or gas, where consequently the people have no easy way to fulfill their everyday energy needs. This has led to many efforts to build solar cookers, water heaters, and household heating systems.

Harry E. Thomason is typical of those who are convinced the sun's energy can be trapped and used to cut down or do away with household heating bills. He collected the sun's heat that fell on the roof of his home and used it almost totally in place of the standby oil furnace.

To trap the sun's energy, Thomason covered the roof of his house with insulation and then put corrugated aluminum on top of the insulation. Water was pumped to the top of the roof and then allowed to trickle down the valleys in the corrugated aluminum. This water was then recirculated to the top of the roof enough times to absorb the sun's energy. The temperature of the water would easily reach 120° F. or even higher.

It was no problem to use this heated water to warm his house during bright, sunny days. To store enough heat to see him through nights and periods of cloudy weather, he built a heat storage tank or bin 7 by 10 by 25 feet. To build this storage bin, he reinforced the basement floor with concrete building blocks and then installed a 1,600-gallon steel tank. The tank was lined with insulation, with lumber to protect the insulation. After fitting the insulated steel tank with pipes to take the heated water in and out, he made it water- and airtight.

Thomason built another space completely surrounding the tank and filled the space with 50 tons of fist-size stones. He argued that the heated water from the roof could be stored in the tank, which in turn would heat the stones. By forcing air through the spaces among the stones and then circulating this warm air through the house, he could harvest the stored heat as needed.

The September 26, 1971, Washington *Post* reported that this solar-heating installation was quite successful. Thomason kept his house at 72° F. during the day and 68° F. at night. His oil bill for a whole year was only $4.65 when his neighbors were paying heating bills well above $200. The system cost approximately $2,500 to build. It was a good investment, and Thomason has gone on to build several more solar-heated homes.

In contrast to this do-it-yourself approach to harvesting the sun's energy, Dr. Peter E. Glaser has designed a much more sophisticated solar-energy-capturing project. Dr. Glaser is a mechanical engineer with Arthur D. Little, Inc., one of the nation's most active research organizations. His idea is to establish a space power station to collect the sun's energy and then beam it to a distribution center here on earth.

Dr. Glaser's plans call for two satellites to be located in space so that at least one is illuminated by the sun at all times, and both have a direct line of sight to the same point on earth. He figures that two satellites would be needed—7,900 miles apart, at an altitude of 22,300 miles, in an orbit parallel to the earth's equatorial plane, traveling east to west.

He has also worked out the dimensions of the solar-energy-collecting surface needed to provide specified amounts of energy. For example, a collecting disk 3.3 miles in diameter would collect enough energy to supply the power requirements of a large fraction of the northeastern United States. Such a collector would weigh 330,000 pounds, plus the weight of its supporting structure.

To send the collected energy to the earth, a system of klystron traveling-wave amplifiers would change the collected energy to microwave form. This would avoid the energy loss that occurs when ordinary sunlight passes through the earth's atmosphere. To transmit the microwave energy, however, calls for an antenna 1.86 miles in diameter at the space power station and a receiving antenna on earth of the same diameter. And there would be problems if anything flew into the beam of microwave energy on its way to the earth from the satellite. Dr. Glaser looks upon the problem of devising the required safety devices and regulations as no more difficult than highway- and air-traffic control. Furthermore, although the power densities in the microwave beam might damage aircraft or injure living tissues, he believes they would not cause major destruction.

To some people, a space solar-energy power station belongs more in the world of science fiction than in the hard, cold facts of the earth's energy resources. Dr. Glaser's proposed satellites, for example, are at least 10,-000 times larger than the largest of today's space power devices. But modern research is rapidly changing the whole field of space exploration. A few years ago it took 100 pounds of instruments to harvest one kilowatt of solar energy; today, only 14 pounds. And instruments are in sight that produce one kilowatt for each 8 pounds of instrumentation.

Present-day solar-energy-collecting cells are about 11 percent efficient, but research suggests that this will soon rise to 18 to 20 percent. It may be that tomorrow's solar-energy-collecting cells will be very lightweight, highly efficient blankets in which the solar cells are laminated between printed circuit sheets of plastic.

To transmit the collected energy from the space station to the earth will call for extremely precise engineering. The receiving antenna on earth would have to be locked into the microwave transmitter in such a way as to adjust for all the forces that act on satellites. These forces include the pressure of other sources of radiation that might hit the power-station satellite, as well as variations in the earth and sun's gravitational forces. The whole idea of space power stations assumes that engineering research will move rapidly toward solving the problems that must be solved before man can commute between the earth and space platforms. If it staggers one's mind to think of space buses and space trucks or tugs taking off on regularly scheduled runs to take people and supplies to and from space power stations, we should remember that there was a day when the idea of commuting people and supplies to ships in the middle of the ocean would have been equally staggering.

It may be that other sources of energy will be developed to take the place of the earth's dwindling supplies of fossil fuels and that it will never be necessary to solve the problems of space power stations. Maybe not. In either case, it is reassuring to know that there are some who have the courage to think and plan for the solution to problems as seemingly overwhelming as those standing between our everyday energy needs and space power stations.

Farrington Daniels of the Solar Energy Laboratory at the University of Wisconsin is one of those who have worked to find better ways to collect and use the sun's energy. In his book, *Direct Use of the Sun's Energy*, he describes solar-energy devices that solve many everyday problems. Professor Daniels was especially interested in helping people in the less industrialized countries of the world increase their energy resources and thereby improve their ways of living. He called this the immediate need for solar-energy research in contrast to the need for long-range research that looks ahead to diminishing supplies of other energy sources.

Professor Daniels describes several different kinds of solar cookers, some of which can be built out of materials readily available in underdeveloped countries or can be obtained at little cost. He tells how solar energy can be used to heat water, heat homes, dry farm crops, and produce fresh water by the distillation of seawater. Although more expensive and complicated devices are involved, he also describes solar-energy cooling systems and electric generators.

Professor Daniels enthusiastically promoted research

in the collection and use of the sun's energy. He knew that solar energy couldn't compete economically where other energy resources were abundant. He was especially concerned about people in the sun-rich, fuel-poor areas of the world. Perhaps research that could help these people right now will be just as much appreciated in the future by people in different parts of the world.

3 Natural Gas

Before 1820, natural gas was not much more than a puzzling curiosity. In several parts of the world, "burning springs" added a ghostly effect to the nighttime landscape. Night after night, blue and yellow flames would play among the rocks and ripples of these mysterious springs. The travel stories written by the early explorers of America describe these curiosities. George Washington, for example, was so fascinated by a burning spring he found in the Kanawha River Valley near present-day Charleston, West Virginia, that he purchased the surrounding land.

It took many years, however, for people to learn how to take advantage of this natural energy resource. During the 1820's, a particularly inventive young man, William A. Hart, found a way to use the gas from a burning spring near a village in the northwest corner of New York State. He managed to capture the flammable gas before it ignited or escaped into the air and then used pipes to take it to where he found a use for it. His success is reported in newspaper stories, which say that several buildings in Fredonia, New York, were lit with gas as early as 1825.

Hart and his business partners convinced the United States government to use gas to replace the oil-fueled lamp that served a lighthouse on nearby Lake Erie. Wood pipes were used to take the gas from a nearby burning spring to keep the gas-fueled lamp burning for more than twenty-five years. The gas-burning lighthouse has been restored as a tourist attraction and a memorial to the pioneering spirit of William A. Hart.

The discovery and exploitation of natural gas as an energy resource include a particularly disturbing state of affairs. Too much of the earth's store of natural gas was badly managed or downright wasted before people learned how to use it. Once the value of natural gas as an energy resource was realized and the problems involved in transporting it were solved, an enormous market developed rapidly—a market that threatens to consume the remaining supply within a very few years.

The problems associated with not knowing what to do with natural gas really mushroomed during the second half of the nineteenth century. Starting with the first oil well ever drilled in the United States at Titusville, Pennsylvania, in 1859, a large industry developed; its business was to harvest, refine, and distribute petroleum products, especially kerosene, or "lamp oil." To show how rapidly this industry grew, enough oil wells had been drilled by 1870 to produce 5,000,000 barrels of oil each year. By 1880 production had jumped to nearly 30,000,-

A number of specially designed tankers like this one transport natural gas in liquefied form at minus 260 degrees Fahrenheit. Liquefied natural gas takes up only 1/600th the volume compared to its gaseous state. (Courtesy of Brooklyn Union Gas Company)

000 barrels, to 50,000,000 by 1890, and to 60,000,000 by 1900.

With the coming of the automobile, particularly during the years between 1910 and 1920, a huge new market for another petroleum product appeared. In response to this market, by 1920 enough oil wells had been drilled in the United States to produce more than 400,000,000 barrels of oil each year, and by 1930, production had increased to more than 900,000,000 barrels.

The Energy Resource That Was Opened Up Too Soon

In many oil fields the rock formations that yield oil also contain gas or are close to formations that do. When oil wells were drilled, the oil industry often opened up a supply of gas at the same time. But gas is much more difficult to package or transport than oil or oil products. When a gas well comes in, either the gas must be transported directly to where it will be used or the well must be capped to prevent the gas from leaving its underground-storage rock formations.

In recent years the gas industry has found ways to store gas with one interesting solution to the problem taking advantage of previously depleted gas fields. Where a depleted gas field is located near a large market for natural gas, a supply of gas is pumped from more distant fields and forced back down into the original gas reservoir. The gas can be stored during the periods of the year when the demand is low. At first thought, to store gas in a hole in the ground doesn't seem very practical. On second thought, because the original store of gas in the reservoir had been kept bottled up for millions of years, we can understand why this solution to the gas storage problem makes sense.

Pound for pound, gas occupies much more volume than oil, coal, or any other liquid or solid fuel. This is why storage is such a difficult and expensive problem. Actually the natural gas that flowed from oil wells during the early days of the petroleum industry didn't become a significant energy resource until the steel industry learned how to make and mass-produce steel pipes.

William Hart used hollowed out logs to carry gas from a burning spring to his gas-fueled lights. Some time later these wooden pipes were replaced with pipes made of lead. But it took many years before the iron and steel industry was ready to make steel pipes that wouldn't leak and were large enough, strong enough, and cheap enough to solve the gas-transport problem. The industry solved these problems and steel pipe became generally available during the 1920's and '30's.

Before 1920, unless a gas well was located near a large city or near an industry that could use this kind of fuel, the well was abandoned. In some cases, before the drillers moved to another location where they hoped to find oil, they sold the well to whoever owned the surrounding land or to a local gas company. By 1920, such markets were using one trillion cubic feet of gas each year.

A gas boom developed quickly, however, when it became possible to build long-distance pipelines. Known

59

gas fields were developed rapidly—but, in retrospect, without much regard to how long they would continue to produce. The author remembers, for example, that thirteen wells were drilled on the southern Ohio farm where he lived as a boy. Actually the total supply of gas in the rocks underlying this one farm could very probably have been harvested by a single well. But the offset rule applied at that time: For each well drilled on one farm, an offset well had to be drilled on neighboring farms. To ensure that all landowners received their share of the gas being harvested, dozens of wells were rapidly drilled in this one relatively small gas field. It was only natural that the total supply of gas was exhausted in a few years.

The amount of gas used in the United States doubled between 1920 and 1930. With the construction of longer and longer pipelines the amount of gas burned throughout the country doubled again by 1945; and again by 1950; and again by 1960. By 1970 nearly 22 trillion cubic feet of gas was being burned each year in the United States. Nearly 50,000,000 homes were cooking, heating, or lighting with gas. So were 3,300,000 commercial and industrial buildings. Consumers' yearly gas bills added to more than $10 billion.

From Too Much to Too Little

Before the 1970's, few people seemed to worry that the earth's supply of gas might be somewhat less than infinite. Advertisements urged people to find more and

60

This is a coal gasification pilot plant to produce pipeline synthetic natural gas. Commercial development of this virtually pollution-free process could substantially increase gas supply in coming years. (Courtesy of Institute of Gas Technology)

more ways to use more and more of this cheap, clean, and convenient source of energy. Everyone was urged to convert from messy, smelly, and inconvenient oil or coal. People acted as though they believed the earth's rocks would create an infinite supply of natural gas, that no matter how many holes were drilled into the earth's rocks, gas would flow forth in abundance.

Thousands of miles of pipelines were strung from the rich gas wells of Texas, Louisiana, Oklahoma, New Mexico, and Kansas to the gas-hungry markets on the East and West coasts. By 1970 more than 225,000 miles of pipeline, some as much as 30 inches in diameter, had been strung across the country. If we include the smaller pipelines that carry gas from user to user, there are more than 900,000 miles of pipeline buried throughout the United States. Gas companies were spending $2.5 billion each year to build additional pipelines and the pumping equipment needed to provide gas for everyone who wanted it.

By 1972 the story of natural gas as an energy resource began taking on a different tone. No longer were gas companies seeking additional customers. No longer did advertisements urge people to convert their oil-burning furnaces to gas, to install gas-fueled rather than electrically powered air conditioners. No longer were industries and power-generating systems urged to stop burning coal and convert to the more convenient, less polluting natural gas. In fact, builders and new homeowners were now told that local gas companies could not take on new customers and that the gas supply was so low they were not sure they could continue to supply even their established customers.

The advertisements sponsored by gas companies carried a new message. One full-page advertisement was titled "The Gas Shortage: What You Can Do About It!" According to this advertisement, "The national energy crisis is the most serious problem facing America today. Without energy supplies, there is no way to solve most other national problems. The worsening crisis involves all primary fuels—oil, gas, coal and nuclear—together with the electricity dependent on those fuels. Natural gas which supplies almost one-third of all U.S. energy needs is already unable to meet current demands."

The advertisement spelled out a definite action program for what should be done about the gas shortage. Gas companies should be permitted to charge higher prices to make more money available to pay for exploratory wells that might discover new fields. More publicly owned land should be opened up for exploratory well drilling, especially the continental shelf off the Atlantic Coast. The pipelines needed to bring oil from the newly found deposits under the North Slope of Alaska should be built promptly, because gas associated with the oil cannot be harvested until the oil is pumped from the storage areas. More natural gas should be imported. The government should put up more money for research on improved methods of making artificial gas from coal.

Another advertisement had the headline: "Deep Gas Wells Are Multimillion Dollar Risks That Must Be Taken." The accompanying message said, "America needs the energy to grow. Experts agree that there are substantial reserves of gas down in the virtually unexplored depths of the Appalachian basin (near the area with a serious supply problem, the Mid-Atlantic States). At this point in time, only about 11 per cent of the gas-bearing formations have been explored. Getting at the large-volume, large-capacity reservoirs means drilling to depths of 20,000 feet and greater. The costs involved with deep drilling are tremendous. A single deep well may cost $3 million. Extended drilling time is a major factor in this increase. It can take as long as nine months to drill to 20,000 feet."

How Much Gas Is There?

Estimates of the amount of gas stored in the rocks under the United States range between approximately 280 and 1,180 trillion cubic feet. The lower figure is based on proved reserves—that is, it is known where this much gas is, and there are good reasons to believe it can be harvested. The higher figure is based on unproved, or potential reserves—that is, geological formations exist in the United States that are likely to contain this much gas and, once located, it can be harvested.

The whole earth's proved reserves of natural gas are estimated to be 1,624 trillion cubic feet. Because the geology of other countries is not so well known as that of the United States, it is almost impossible to come up with an accurate estimate of unproved, or potential, reserves. Published estimates range from 7,500 to 30,000 trillion cubic feet. Whether the estimate took into account the possibility that some of this gas can never be recovered explains part of the difference between these figures.

Suppose the use of gas in the United States continues to increase so that over the next ten years the average yearly use becomes 28 trillion cubic feet. On this basis, all the proved reserves of natural gas in the United States will have been used up by 1982. If the estimates of unproved, or potential, reserves turn out to be reasonably accurate, and if all this gas can be gotten out of storage rocks and into pipelines, the additional gas would last another forty years.

It is difficult for some people to believe that a critical shortage of natural gas is really this close at hand. People who tend to be particularly suspicious question the accuracy of estimates of gas reserves as well as the rate at which gas is being burned. They think a gas shortage is being played up to help gas companies make more money—by charging higher prices, or by getting government assistance through tax advantages or other benefits.

F. Donald Hart, president of the American Gas Association, recently replied to these ideas in a "letters to the editor" column of a national newspaper. He denied that the crisis is being artificially created, and he agreed that "there is no shortage of natural gas under the ground." He argued that "the key to the problem is to make it worthwhile for producers of gas to invest the tremendous sums of money and take the significant risks involved in drilling for new supplies of natural gas."

The most thought-provoking part of the argument over the reality of the gas shortage is that all parties seem to agree that in about fifty years, the gas fields of the United States will be totally depleted in terms of both proved and potential reserves.

The Origin of Natural Gas

The creation of the earth's store of natural gas is a one-time event. It happened as many as a billion years ago, when large areas of the earth were covered with dense swamps. For thousands or even millions of years, the remains of dead plants collected at the bottoms of these swamps or were washed into nearby lakes or oceans. When the earth's crust shifted during prehistoric earthquakes and other earth movements, these deposits of dead plant material were covered with thick layers of mud and sand. In time, chemistry changed the decaying plant material into totally new substances, including gas, oil, and coal. And the overlying layers of mud and sand became beds of shale, limestone, sandstone, and other sedimentary rocks.

In some situations, the overlying rocks formed a gas-tight trap that kept the accumulating gas and oil from escaping. Volcanic action sometimes bulged the earth's crust so as to form a gas-tight dome. Or the rock layers slid and slanted in such a way as to form huge pockets that kept the newly forming gas from slowly bubbling away. Pockets where the gas was trapped have become today's gas fields.

The earth's deposits of gas, oil, and coal may originally have been produced in the same rock formations. But gas can move through or among some kinds of rock formations where oil cannot. Consequently deposits of gas are not always found where there are deposits of oil. An estimated 80 percent of the gas that remains under the United States, for example, is not associated with or dissolved in accompanying oil deposits. On this basis, many yet undiscovered gas fields will be difficult to locate.

Attempts to Provide More Gas

The threat of an increasingly serious shortage of gas disturbs many people. The millions of people whose homes and businesses are fueled with gas foresee the expense and inconvenience of shifting to some other energy resource. An inadequate supply of gas is an exceedingly serious threat to the multibillion-dollar industry in the business of recovering and distributing gas—not to mention all the other business interests that make and sell gas-fueled appliances. These fears and threats are driving people to seek ways to increase the gas supply.

What can be done to head off the troubles that would surely come if and when our gas pipelines run dry? Deeper wells can be drilled into more of the earth's rocks in the hope of finding new gas fields. But there is a kind of uneasiness about this approach to solving the problem. The fact remains that someday all the earth's store of natural gas will have been discovered, recovered, and burned. The showdown may be postponed a few years, but that is all.

Another approach seeks to improve the methods for

recovering a greater portion of the gas that exists in underground deposits. There is a government-sponsored research effort, for example, that attempts to jar loose the gas trapped in hard-to-get-to formations by setting off underground nuclear blasts. Officials estimate that as much as 317 trillion cubic feet of additional gas could be obtained from the rocks under the United States if selected formations were fractured by nuclear blasts.

This is a highly controversial project. Opponents argue that the gas that would be harvested and pumped into cities and towns would be radioactive. More opposition comes from businessmen who hope to extract oil from oil-bearing rocks near or on the surface of large areas in the Rocky Mountain region. They are afraid that nuclear blasts would destroy the oil shale or otherwise interfere with the plans to develop this energy resource. They also argue that the amount of oil from the oil shale would yield more than 100 times as much energy as the gas jarred loose by underground blasts.

Another solution to the gas shortage in the United States depends on importing gas from other countries. Here again, packaging is the biggest difficulty. The gas is liquefied so that as much gas as possible can be shipped in manageable containers. If gas is cooled approximately 290 Fahrenheit degrees below the freezing temperature of water, it becomes liquid. When gas liquefies, it takes up much less space. For example, 600 cubic feet of room-temperature gas will fit into 1 cubic foot of space when it is liquefied. By liquefying the gas, a ship can haul enough to make the business economically worthwhile.

It is expensive, however, to liquefy gas. Pumps are needed to put the gas under pressure. To cool the gas calls for costly equipment. And the ships that haul liquefied gas must be equipped with special refrigerating, or cryogenic, apparatus in order to maintain the gas in liquid condition. For these reasons imported gas can cost three or four times as much as locally produced gas.

Other problems arise when one nation must import its energy resources from another. In 1969, for example, Algeria agreed to sell one billion cubic feet of gas per day to a group of U.S. gas companies over a twenty-five-year period beginning in 1976. The agreed price was 30.5 cents per thousand cubic feet. All parties to the agreement realized they would have to invest $1.6 billion in plants to liquefy the gas, to build ships to transport it, to construct terminal facilities to move it to and from the ships, and to string pipelines to get it to markets.

Having entered into the agreement, the U.S. gas companies went ahead to establish markets for the gas they expected to import. Many of these markets were in Eastern cities, where there were good indications that available supplies would soon be inadequate. In fact, by 1971 the domestic supply-and-demand situation prompted the gas companies to ask Algeria to provide double the amount of gas originally agreed upon. By this time, however, Algeria found that gas could be sold to European countries for 42 cents per thousand cubic feet. This increase in the going price for gas caused the original agreement with the U.S. gas companies to come up for renegotiation.

In the meantime, the U.S. government had set legal ceilings on the price the gas companies could charge their customers. These ceiling prices, in effect, kept the importing companies from meeting the new price demands of Algeria. To add to the confusion, much of the $1.6 billion investment needed to get the original deal under way was to be provided by New York bankers. These financiers threatened to withdraw their support unless they were sure the deal would go through.

All in all, to find and harvest gas from the earth's rocks poses certain kinds of problems. To import gas from other nations invites all kinds of additional problems, especially all the complex problems involved in international relations.

Similarities in the origins of gas and coal suggest that it is possible to make gas from coal. Furthermore, the earth's rocks contain enough coal to continue to meet the demand for several hundred years. The chemistry of making gas from coal is relatively simple. Coal consists of approximately 75 percent carbon. Gas, if it is pure methane, consists of 75 percent carbon and 25 percent hydrogen. Water is a cheap source of hydrogen. If water is sprayed onto a bed of hot, burning coal, hydrogen and carbon monoxide are produced. If the mixture of these two gases is passed over a nickel catalyst while they are still hot, a new reaction occurs, and methane and water are produced. When the mixture of methane and water is cooled, the water condenses and can be drained away, leaving the methane to be sold as artificial rather than natural gas.

Again, however, this is an expensive operation. Much of the original energy in the coal doesn't end up in the manufactured gas. And the coal cannot be burned in ordinary air. Air contains four times as much nitrogen as oxygen. Unless pure oxygen is used when the coal is burned, the resulting methane would be diluted with nitrogen. Nitrogen doesn't burn, and its presence in the marketed gas would lead to many problems, including air pollution.

Methods for making gas from coal have been known for a long time. Before the steel industry solved the problems of making steel pipe, many cities had gas works. But the neighborhood of the gas works invariably had a reputation as the worst part of town in which to live or work. The air was almost always badly polluted with fumes. Consequently artificial-gas plants were put out of business as soon as it became possible to pipe gas from distant sources.

Processes for converting coal to gas are being improved. In one of the improved methods, the coal is ground to a fine powder. A blast of very hot air is used to mix the powdered coal with a light fuel oil to form a thin mud or slurry. The slurry, in turn, is sprayed into a reaction chamber that contains a bed of hot coals. The gases produced in the reaction chamber contain a high yield of methane. The unreacted coal, oil, and steam are removed by cooling and washing. Additional scrubbing processes remove other impurities, especially the sulfur oxides that are formed if the original coal contained sulfur, as most coal deposits do.

Manufactured gas is more expensive than natural gas. It is reassuring to know, however, that gaseous fuel will be available so long as our coal resources last—particularly because certain industrial processes and products are almost totally dependent on gas as their primary energy resource.

Natural Gas and the "Energy Binge"

It is reassuring to know that coal can be used to piece out the earth's supply of natural gas. But not totally so. Civilization has been a long and, we hope, a continuing story. Everything man has done in past years called for one or another source of energy. All that man will add to the progress of civilization will call for one or another source of energy. And the future can be a long, long stretch of time. True, for each of us or for any one generation, the future may stretch no farther than the traditional "three score and ten" years. But should one generation, particularly our generation, do something or fail to do something that can make an impact on the whole continuing stream of civilization?

The earth's store of fossil fuels, its deposits of gas, oil, and coal, can be thought of as an energy resource tucked away along the mainstream of man's efforts to build a civilization. For thousands of years, man was not aware of or at least paid little attention to this store of energy. Achievements depended only upon the energy that came streaming into the world from the sun. This was sufficient energy to let man build cathedrals and castles, to

72

create paintings and sculptures, to power printing presses, spinning mills and foundries, factories and farms, to compose and perform music and drama, and to drive carriages or sail ships.

Yes, the stream of civilization was pretty well established before man found occasion to use the energy stored in gas, oil, and coal. But there is the question of the scope and speed of this stream of activities and achievements. The input of the sun's energy allowed man to grow enough grain and other farm crops to support only so many people. The forests of the world produced only enough timber and firewood to meet the needs of only so many people. The same situation applied to the windmills and waterwheels that ground the grain, spun the cloth, or powered the machines that made civilization possible.

The discovery of fossil fuels tucked away, as it were, along the stream of civilization allowed people to build whatever they wanted to build, to go wherever they wanted to go. In the original metaphor of this book the whole earth toy began to spin at an ever-increasing speed. In time whole generations of people became accustomed to civilization moving swiftly, and they may have lost sight of the fact that all this was possible only because the earth's energy resources were being used up far more rapidly than at any other time in the history of civilization.

Natural gas is one of these energy resources. It may be the first of the earth's energy resources to be totally consumed. If the nation's gas pipelines were to go dry, millions of people would have to be provided with new sources of energy. If not, these people would have to make serious adjustments in the way they live. Their part of the whole earth toy would surely slow down.

These adjustments are inevitable. The only question is how soon they will have to be made. Whether or not the adjustments will put millions of people to serious inconvenience can well depend upon what action is taken during the closing years of the twentieth century. In the words of an advertisement sponsored by the nation's gas companies, "You owe it to yourself and your community to become more informed about the grave national energy situation." Better-informed people will be more likely to come up with intelligent solutions.

4 Energy from Petroleum

Filling stations, oil trucks, and refineries with their mazes of pipelines and storage tanks are very much a part of today's landscape. At every airport, huge tanker trucks keep gasoline or jet fuel flowing into waiting planes. At every harbor there are dockside "filling stations," or tanker ships, to pump fuel into the boats and ships that travel the waterways of the world. Millions of homes and public and commercial buildings are warmed in winter and cooled in summer by burning the fuel pumped into furnaces and air-conditioning units.

A hundred years ago it was a different story. What will be the situation a hundred years from now? This question cannot be answered. It is not that geologists and petroleum engineers are unable to estimate how much oil there is in the earth's rocks or that nobody knows how much oil is consumed each year. We can only guess about the changes that will occur in man's oil-consuming activities in the future because we don't know what will be done when oil becomes increasingly scarce.

And become increasingly scarce it will. The earth's

store of oil is a one-time thing. It is highly unlikely that the underground deposits of petroleum are being or will ever be replenished. Every drop of oil burned today or yesterday or tomorrow has been taken from nature's oil-making processes, and in general, these processes went out of business 500,000,000 years ago.

Oil as a Fossil Fuel

The energy from burning gasoline and other petroleum products was put into these fuels when the sun's heat and light fell on swamps and bogs millions of years ago. This energy was stored because of a fortunate set of circumstances. Usually plants absorb the sun's energy; the plants grow; then they die and decay. The energy built into their body tissues either escapes or is temporarily transferred to the body of some other living system.

How the plant materials in those prehistoric swamps were converted to fossil fuels remains pretty much a mystery. We do know that these fuel deposits have been buried under great thicknesses of shale, limestone, sandstone, and other sedimentary rocks. This suggests that where the swamps existed, the earth's crust has undergone spectacular changes. Areas once well above sea level became submerged. And where there are several different levels of fossil-fuel-bearing rocks in the same area, there must have been times when the earth's crust rose in one section and fell in another.

When a submerging swamp became covered with in-

An oil well fire near Canton, Oklahoma. (Courtesy of Bureau of Mines, U.S. Department of the Interior)

creasing depths of water, the plant remains would be covered with muds, limes, sands, and other sediment from the erosion of newly emerging areas. Apparently the plant material was covered under circumstances that prevented the usual decay of organic material. Under these circumstances the complex, energy-rich molecules which make up plant tissues were not broken down to carbon dioxide and water and therefore did not lose their stored energy.

Thick layers of deeply buried plant material became huge chemical factories. Partial decay together with the pressure of the overlying rock layers produced high temperatures. These conditions favored the chemical reactions necessary to convert the plant roots, stems, and leaves to the mixtures of chemical compounds that form oil, gas, and coal. Because the compounds retain some of the sun's energy that had originally fallen on the prehistoric swamps, these reactions have produced valuable energy resources. It takes only the strike of a match or the spark from a spark plug to release this trapped energy.

But here's the catch. The circumstances that produced the whole earth's store of oil no longer exist. There is little reason to believe that new deposits of oil are being formed in some hidden place or undiscovered way. Even if they are, remember that production of today's store of oil began 500,000,000 years ago, and it could take that long to replace it.

Unless something happens to change current practices, the earth's 500,000,000-year-old store of oil will have been discovered and burned in only a fraction of man's total history. Oil wasn't burned in appreciable quantities before about 1857. But how rapidly the usefulness of this energy resource has increased is suggested by the fact that as much oil was burned during the ten years between 1957 and 1967 as was burned between 1857 and 1957. Such facts compel us to face the question: How long will the earth's store of oil last?

How Much Oil Is There?

How long the earth's store of oil will last depends on two things: how rapidly we are consuming oil and how much oil can be gotten from the earth's rocks. We know how much oil is being consumed each year throughout the world. It is not easy, however, to estimate accurately how much oil there is in the earth's oil-bearing rocks. And for several reasons. Before the amount of oil in any storage area can be determined, the storage area must be located. This calls for exploratory, or wildcat, wells. To drill holes several miles into the earth's crust is expensive, and large portions of the earth have not yet been explored. It is almost impossible, or at least economically impractical to harvest all the oil in any storage area. Consequently, to estimate the amount of oil that can be gotten from the whole earth's oil-storage areas requires the estimator to guess about the availability of the oil that is known to exist in already-explored areas.

Attempts to estimate the worldwide reserves of petroleum provide interesting lessons on how man tends to react to issues and problems. Different people have come up with widely different estimates of how much oil there is in the earth's rocks. They also differ widely in their attitudes about the threat of an impending shortage of oil. Some say there will always be enough oil; there always has been and there always will be. Others feel that within a few years, twenty or so, the available supply of oil will have shrunk so that the price of gasoline and other petroleum products will be so high that few people will be able to afford them. The threat of an oil shortage is so worrisome to these people that they can see their whole life-style or way of making a living go up in smoke —the smoke from burning the last few barrels of oil from the earth's storehouse. Other people seem to accept the idea that we are rapidly consuming the earth's supply of oil but aren't especially worried because they are confident that other energy resources will be found long before we run out of oil.

Because of the importance of petroleum products in the United States and the seriousness of the planning that must anticipate an oil shortage, there has been great effort to estimate the amount of oil that can be gotten from the rocks under this nation. A recent summary lists more than a dozen estimates that have been published by recognized authorities. These estimates range from a low of 110 billion barrels to a high of 590 billion barrels. The average of fourteen estimates is 270 billion barrels; but there are sharp arguments about what this average really means, whether the estimate closest to the average is any more likely to provide the "correct" answer.

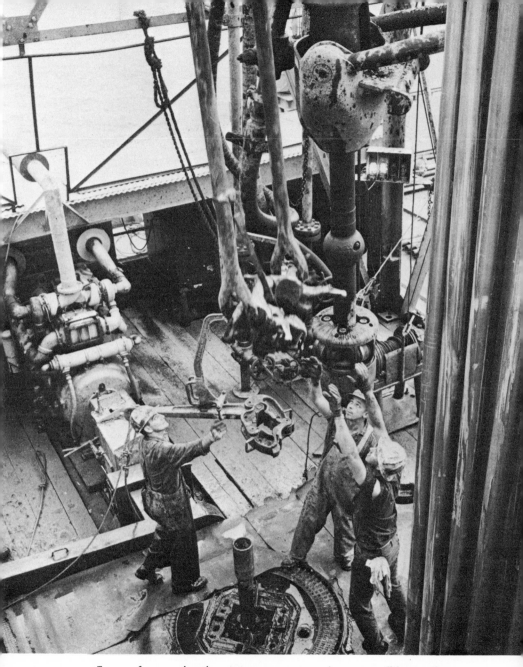

Energy from under the sea: a crew at work on an offshore oil drilling station. (Courtesy of Texaco, Inc.)

There is less disagreement about how much oil is consumed each year in the United States. Estimates for 1970 cluster rather closely around 5 billion barrels. It is also generally agreed that the rate of consumption increases approximately 7 percent each year. This means that annual consumption doubles every ten years. On this basis, assuming that the current rate of increase continues, consumption will rise to 10 billion barrels by 1980, to 20 billion barrels by 1990, and so on.

Not many years remain to plan what must be done to head off the problems that would become drastic if the nation found itself without adequate supplies of oil. This planning could be done much better if truly accurate estimates of the remaining oil supply were available. When we realize how oil deposits were formed, however, it becomes clear why such estimates are not at all easy to obtain.

Estimating Oil Reserves

Certainly, no oil will ever be found in those parts of the world where the earth's crust is made of igneous or metamorphic rather than sedimentary rocks. In eastern Canada, in Norway and Sweden, and throughout a large part of Africa, for example, there are no thick layers of sedimentary rocks. There can be no oil deposits in these areas.

Where sedimentary rocks make up the earth's crust, there is at least a possibility that they contain oil, assuming, of course, that there were large prehistoric swamps in the area. For rock layers to trap and store oil for millions of years, however, requires a definite set of conditions. The rocks must be sufficiently porous to allow them to absorb oil. Oil-bearing formations must be arranged in a way that prevents water from seeping in and displacing the less dense oil.

One type of rock formation that can trap and store oil deposits is something like an upside-down shallow pan or basin whose bottom forms an oil-tight dome or ceiling. These basins may be only a few hundred feet in diameter or as large as twenty or more miles. The oil they contain varies accordingly. In another type of oil-bearing formation, gently sloping layers of porous rock jut up against a more steeply tilted layer of oil-tight rocks. If the oil-bearing layers are also covered by an oil-tight rock layer, the oil is trapped and stored.

The kinds and arrangement of the rock layers under the United States have been pretty well mapped. Consequently the locations of the major oil deposits in the United States are known. This doesn't mean, however, that there aren't undiscovered deposits or undiscovered pools in the major oil-bearing formations. Nor is it known for sure how much oil can be gotten from the known oil fields. It isn't that people don't want to obtain this information. They do. This is one way to become a millionaire or even a billionaire.

Highly trained geologists spend their lives looking for clues that will lead them to undiscovered oil deposits. Businessmen in the oil industry encourage engineers and scientists to pool their knowledge and design instruments that will accurately map underground rock layers. But when those whose business it is to locate oil deposits are asked how they go about finding oil, they are likely to shrug their shoulders and say, "Oil is where you find it!"

The expense of drilling a new hole through the earth's rocks in the hope of finding an oil deposit is a very risky investment. In fact, speculative drilling is called wildcatting, and according to a recent article in the *Wall Street Journal,* few successful drillers sink much of their own money into oil wells, which is one big reason they are successful. The article reports that the risks of wildcatting are spread among a large number of investors—an attractive scheme, considering the enormous income that such an investment will produce if a valuable oil pool is discovered.

To add to the attractiveness of the drilling-fund investment scheme, it is understood that even if no new oil fields are discovered, the investor can recover the investment through an income-tax write-off. To show how this works, suppose an investor is already rich enough to have 50 percent of his income go for income tax. He invests, say, $50,000 in a drilling fund. The managers of the fund have a wildcat well drilled, but it turns out to be a dry hole. The total cost of the dry hole was many times the one investor's $50,000. Consequently the investor now becomes responsible for his share of the total cost of the exploratory well. This can mean that rather than becoming rich, the investor has now "lost" $110,000, which includes what the managers of the investment fund had "lent" the investor. On this basis, the investor can deduct the $110,000 loss from his other net earnings and incomes when it comes time to pay his income tax. If the investor is in the 50 percent income-tax bracket, he "saves" $55,000 that he would have had to pay on his other income. And this is $5,000 more than he "lost" on his original $50,000 investment in the drilling fund.

One gets the feeling that there is a bit of financial hanky-panky in this scheme—and especially so when the author of the *Wall Street Journal* article says, "Because the tax writeoffs exist even if the oil doesn't, some funds are suspected of drilling 'dusters' deliberately." If this is the case, it is another reason why it is so difficult to estimate accurately how much oil remains in the rocks under the United States.

Reliable methods for estimating undiscovered oil resources use several kinds of data. Some of these data take advantage of the records kept by the government and the oil industry. It is known, for example, how many feet of newly drilled or wildcat wells have been drilled each year for many past years. It is also known how many barrels of new oil have been discovered during these years. These facts establish an index, barrels of new oil per foot of new drilling, and from this index experts can estimate how much oil will be discovered through future exploratory drilling.

The use of this index is described by M. King Hubbert in the September, 1971, issue of *Scientific American.* Between 1860 and 1920, when oil was fairly easy to find, 194 barrels of new oil were discovered for each foot of exploratory drilling. Between 1920 and 1928, the yield dropped to 167 barrels per foot. Between 1928 and 1938, the ratio rose to 276 barrels per foot, the increase partly accounted for by the discovery of large fields in East Texas and to the development of new geological exploratory techniques. Since 1938 the yield has dropped sharply and has held steady for several years at about 35 barrels of newly discovered oil per foot of exploratory drilling.

Hubbert believes that approximately 50 billion barrels of oil await discovery in the rocks under the United States. This estimate doesn't take into account the 30 to 50 billion barrels believed to exist in the newly discovered fields in Alaska. But it does include the oil resources in offshore or continental shelf areas. He also estimates that approximately 120 billion barrels of oil have already been harvested in the United States.

Obviously Hubbert is only one of those who have tried to estimate how much oil can be gotten from the rocks under the United States. He is more pessimistic than some and more optimistic than others. His work is quoted widely. Some other estimators agree with his methods and conclusions. Others don't. Some of his critics say his estimate of 175 billion barrels as the total amount of oil that has existed or still exists in the United States is more than 50 billion barrels too low. Even so, Hubbert points out, if the current rate of consumption continues, an additional 50 billion barrels would postpone the date of extreme petroleum shortage by only ten years or so.

Unless things change abruptly, the years between 1925 and 2025 could stand out in the history of civilization as the century when the people of the United States consumed more than 80 percent of their nation's oil supply. This enormous rate of energy-resource depletion has made possible one of the world's most technologically affluent societies. But it raises very thought-provoking questions. How will future generations feel when they look back at this one-hundred-year period in the total history of man's development and use of the earth's energy resources? Will the achievements of that wonderful, technologically affluent society justify their cost?

The Oil-Burning Chapter in the History of Civilization

How will future generations feel toward those who hastened the discovery, development, and consumption of the nation's oil resources? Some of these people became very wealthy, and their names stand out in the "oil-burning era" of the nation's history. How will history treat them? Will they receive credit for advancing mankind toward increasingly better standards of living? Will they be blamed for creating a fool's paradise, that could exist only so long as we could harvest the energy accumulated during a 500-million-year period in history?

There are other sticky questions. How will the story of the discovery, production, and consumption of oil in the United States affect the worldwide stream of civilization?

The United States certainly assumed leadership in the use of the earth's store of oil.

Our petroleum resources are being used up much faster than any other country's. How will other nations look upon this? Traditionally the world has been divided into the "have" and "have not" nations, and these terms loom large in discussions of worldwide conflicts. Will the possibility of new categories, "have had," "still have," and "have never had" loom equally large in future international relations? Will "still have" nations be willing to share their oil resources with the "have had" nations?

Although there are less data to work with, Hubbert has also tried to estimate the whole earth's supply of oil. In general, he used the same approach as when he estimated the United States' supply. His 1965 estimate put the whole earth's supply of oil at 1,250 billion barrels, of which 850 billion barrels were in land areas and 400 billion barrels in offshore or continental shelf areas. In 1971, he reported the earth's oil supply is now estimated at between 1,350 and 2,100 billion barrels. Based on these estimates and the current worldwide rate of oil consumption, Hubbert calculates that 80 percent of the whole earth's supply of oil will have been consumed by about 2025.

Some people say Hubbert's estimates are pessimistic, that he underestimates the amount of oil that can be gotten from the earth's rocks or that he overestimates the rate of consumption. Other critics say he is overly optimistic, that there will be a worldwide shortage of petroleum even sooner than he predicts. In either case what happens as the United States faces a dwindling supply of oil could well be a preview of a worldwide situation ten or twenty years later.

The worldwide story of dwindling petroleum resources will probably be more complex than the American story alone. For one reason, the earth's supply of oil is not distributed equally among nations. An estimated one-fourth of the whole earth's oil supply lies under the Middle East and North Africa, one-sixth in the USSR, but scarcely one-sixteenth in the whole continent of South America. North America holds one-sixth of the whole earth's supply of oil and approximately three-fourths of this is in the United States.

The United States consumes almost one-third the oil used each year throughout the world. The United States has already consumed half its original oil supply, while most other nations have used up only a quarter of their oil resources. The United States will have to adjust to dwindling oil supplies before adjustments become urgent in other nations.

Increasing the Oil Supply

Not counting imports, if American oil supplies become inadequate, is there any way to increase the amount of oil that can be gotten from underground storage areas? To answer this question, we become involved with how oil is removed from the rocks in which it is trapped and brought to the surface.

Sometimes in the early days of oil-well drilling, the drill punched through the cap rocks above an oil-bearing rock stratum, and large quantities of oil gushed up to produce an enormous oil fountain. The force that brought the oil to the surface came from the sheer weight of the overlying rocks. Furthermore, the conditions that produce oil are such that natural gas is often stored with it. When a hole is punched into a rock formation where oil and gas are trapped together, gas from the surrounding area rushes toward the hole and drives the oil with it.

Sometimes the oil that has accumulated in a dome-shaped reservoir is floating on the water or mixed with it. The cap strata that hold the oil underground also trap the water. If a hole is drilled through the trapping layers of rock, the pressure of the surrounding water forces the oil toward the surface.

Usually the forces that cause gushers play out before all the oil in the reservoir is brought to the surface. When this happens, pumps are used to lift the oil after it accumulates in the bottom of the well. After much of the oil has been taken from the oil-bearing rock layers, there is not enough pressure on the remaining oil to force it to seep into the well. Steps must be taken to force the oil that remains in the surrounding rocks to move toward the well.

One way to do this is to use injection wells. Rather than pump oil from the oil-bearing rock layers, water, gas, or some other cheap fluid is pumped down injection wells to force the oil toward production wells. The injection method for recovering the maximum yield from an oil pool has become highly developed. One modern version uses the gases that first boil off heated petroleum. These gases are good solvents for petroleum, and if they are forced to diffuse through oil-bearing rocks and sands, the oil is flushed out of the rocks and moved toward a production well. To lower the cost of this type of injection process, only a bubble of the gas is pumped down the injection well. A cheaper fluid is then used to back up the gas bubble. As much as 85 percent of the oil in an underground storage pool can be harvested with such a method.

In large oil fields a network of injection wells is planned so as to surround production wells. A favorite design is the five-spot system, which puts one producing well at the center of an imaginary square with injection wells at each of the corners. How closely spaced the wells are depends on the viscosity of the oil being harvested and the kinds of rocks or sands in which it has been trapped. Furthermore, the entire oil-bearing rock layer may have local hills and valleys, which form barriers that affect the flow of oil from the injector to the production wells.

In the early days of oil production it was easier to harvest whatever oil was naturally forced up from its storage areas. When a well ran dry, it was not too difficult to discover a new field. Today new fields are becoming increasingly hard to find, and efforts are being made to find better ways to harvest all the oil in already discovered fields. One new idea that promises to harvest all the oil in a pool depends on burning some of the oil while it is still in its original oil-bearing rocks or sands. If air is forced down an injection well to produce an oil and air mixture in an oil-bearing stratum, the mixture can be ignited. In turn, the heat from burning oil near an injection well will vaporize the surrounding oil, and the resulting gas pressure will force the not-yet-burned oil to flow toward the producing wells. ↓

For each fifteen barrels of oil burned in this type of recovery process, eighty-five may be collected from the producing wells. To add to the attractiveness of the process, the more volatile, more expensive components of petroleum are flushed out first. The cheaper and less volatile components are left behind as fuel for the flushing-out process.

Oil Shales and Tar Sands

In the mountainous country of northwestern Colorado, Utah, and Wyoming are vast amounts of surface rocks that promise to piece out the world's supply of oil. These oil shale beds contain kerogen, and although kerogen cannot be coaxed from shale the way oil is from oil-bearing rocks and sands, heat will do the job. If kerogen-bearing rocks are heated to approximately 1000° F., the kerogen produces oil, gas, and a mineral residue.

The richest deposits yield as much as 100 gallons of oil per ton of shale; the leanest deposits, less than 5 gallons per ton. It is difficult to estimate accurately how much oil could be gotten from the total shale deposit because the yield from different samples varies widely. One estimator believes as much as 80 billion barrels of oil could be harvested from the easily accessible layers of shale and yield at least 25 gallons of oil per ton of shale. Another estimator, who includes shales that yield no more than 10 gallons of oil per ton, believes the total resource contains nearly 1,500 billion gallons of oil. It is believed that other parts of the world contain enough oil shale to add another 1,000 billion barrels to this estimate.

Methods have not yet been developed whereby oil can be gotten from oil shales cheaply enough to compete with oil and gas from wells. Research is at the pilot plant stage. One such plant produces 30 gallons of oil per ton of processed rock. From this yield of raw oil, the refinery obtains 14 gallons of gasoline, 10 gallons of diesel fuel, 0.6 gallon of bottled gas, 0.6 gallon of furnace oil, and 40 pounds of coke.

Serious problems must be solved before the pilot operation can go commercial. One problem is getting rid of the huge amount of waste. To produce 50,000 barrels of oil per day, for example, would pile up 60,000 tons of waste rocks—and it would take nature's soil-making processes a long time to work this waste material into something other than landscape pollution. Assuming these problems will be solved, it is reassuring to know that there is enough oil in these shales to supply the United States for about fifty years.

Kerogen is a strange substance. Geologists don't agree on where it fits into the family of petroleumlike substances. It is certainly a product of the partial decomposition of prehistoric vegetation. But there are many unanswered questions about the conditions that existed while kerogen was being made.

Bitumen is a substance that is closely related to kerogen. It is more like petroleum, however, and is found soaked into beds of sand. The richest deposits of these tar sands occur in Canada, but there are enough deposits throughout North America to yield an estimated 700 billion barrels of oil. It is easier to mine tar sands than to process oil shales, but, ton for ton, the yield of oil is less. To produce 50,000 barrels of tar-sand oil, for example, would require handling well over 100,000 tons of raw sand.

From Crude Oil to Refined Product

When it is first recovered from underground storage areas, petroleum is a crude mixture of many different kinds of carbon-hydrogen compounds. Each oil field yields its own unique blend of the mixture. Different wells in the same field don't always yield identical crude oil, and the same well's output may change as the supply of oil in the storage pool runs out. Different kinds of impurities also show up in different oil fields. Sulfur, for example, occurs in many oil deposits, but there are many that are sulfur-free.

To recall the prehistoric origin of petroleum helps explain why its composition differs in various pools. The kinds and numbers of plants might have varied from swamp to swamp. The environmental conditions that changed the plant material to petroleum might have been just as fickle as today's winds and weather. By no means can we assume that the origins of the earth's store of petroleum were accompanied by carefully controlled chemistry—regardless of how successful this chemistry must have been.

Highly controlled chemistry, however, is used in the refining of petroleum. A modern refinery does more than simply separate the mixture of hydrocarbons into its various components and take out impurities. Cheaper components are converted to compounds or mixtures of compounds for which there is greater demand.

Only two elements, carbon and hydrogen, account for most of the chemistry of petroleum. Differences in the number of carbon and hydrogen atoms and their arrangement in the various components are the important factors.

If we assume carbon atoms have four times the combining capacity (valence) of hydrogen atoms, then the simplest component of petroleum appears to be a molecule consisting of one carbon and four hydrogen atoms. This compound—methane, or CH_4—however, is so volatile that it boils out of petroleum unless the petroleum is kept very cold or under great pressure. The same goes for the next more complex hydrocarbon—ethane, or C_2H_6. Notice that each of the two carbon atoms share one of their valence bonds to hold them together; this leaves only six bonds for the hydrogen atoms.

The three- and four-carbon-atom hydrocarbons—propane, or C_3H_8, and butane, or C_4H_{10}—are pretty much like methane and ethane, but they are less volatile and remain dissolved in crude petroleum at ordinary temperatures and pressures. These gases, however, are rather easily boiled out of petroleum and are bottled under pressure to become the commonly used bottle-gas fuel.

Moving along toward the higher numbers of carbon-atom hydrocarbons, we come to the important mixture that makes up gasoline. Molecules that consist of between five and eleven carbon atoms per molecule are generally suitable for automobile fuel. These hydrocarbons exist as liquids at ordinary temperatures and pressures, but they are sufficiently volatile to be easily vaporized in the carburetors of internal-combustion engines. With the great demand for gasoline, it is easy to see why hydrocarbons with approximately eight carbon atoms per molecule—octane, or C_8H_{18}—make up one of the most demanded and, consequently, most expensive components of petroleum.

Kerosene is a mixture of hydrocarbons with molecules of about ten carbon atoms. Although this mixture of higher-number carbon-atom hydrocarbons is not sufficiently volatile for use in automobiles, kerosene can be used for diesel and jet engines. The specific requirements of any engine can be met by adjusting the ratio between high- and low-number carbon-atom hydrocarbons when the fuel is being blended.

In general the higher the number of carbon atoms per molecule, the lower the volatility. As volatility decreases, so does market value. Furnace oil, for example, is a mixture of hydrocarbons with less volatility than required for vehicle fuel and is one of the cheapest products of the refining industry. The asphalt or coke residue left after all the readily evaporated components have been boiled off is the cheapest of all the industry's products. Lubricating oils and greases consist of higher-number carbon hydrocarbons, but these products require special refining to remove wax, asphalt, or other impurities. This accounts for the higher cost per gallon of these products.

Getting More Gasoline from Crude Oil

There is more to refining crude petroleum than simply boiling off each of the separate components one at a time. Raw petroleum, as it comes from the earth's rocks, may have bits of clay or tiny sand particles suspended in it. More or less water, sulfur, or other impurities may be mixed with or dissolved in the petroleum. All these contaminating substances must be removed.

Another reason why the petroleum-refining industry is a multibillion-dollar endeavor is that crude petroleum doesn't contain the greatest percentage of the most salable products. There is a great demand for gasoline because it is such a convenient fuel for automobiles, and people are willing to pay a good price. Five gallons of crude oil, however, generally yields only one gallon of gasoline. Only one-fifth of the total mix of hydrocarbon molecules have the required number of carbon atoms per molecule to give these fractions of crude oil the necessary properties for auto fuel. But the fuel value of some of these molecules could be improved if their atoms could be rearranged.

97

The refining industry has developed methods whereby small-molecule hydrocarbons can be combined to form larger molecules. And molecules that are too big can be cracked to form smaller molecules. In practice an additional fifth of the original crude oil can be reworked into gasoline by one or another rebuilding process. The small molecules of methane, for example, can be joined to one another or combined with molecules of propane or butane to make good gasoline.

Hydrocarbons with about eight carbon atoms per molecule aren't good automobile fuel unless the carbon atoms are arranged properly. They cannot be in an unbranched chain. To convert this fraction of the crude petroleum to gasoline, the unbranched chain of carbon atoms is broken under circumstances that cause the fragments to rejoin in branched chains. These long-chain hydrocarbons can also be converted to automobile fuel by causing the ends of the chain to join and form a closed ring, or cyclohydrocarbons.

Oil and the Pollution of the Environment

There are many things that petroleum engineers can do to increase the yield of gasoline from crude oil. The processes, however, call for high temperatures, high pressures, and additional chemicals. And they produce large quantities of byproducts and wastes that have to be removed. Many of these wastes are gases or oil-like substances that pollute the air if they escape from smokestacks. If they are dumped into streams or lakes, a different but equally serious form of environmental pollution results—pollution that is painfully familiar to everyone who lives or travels near badly managed oil refineries.

Ideally an oil refinery should obtain its crude oil without allowing any to be spilled before it is fed into the refining processes. Ideally all the refining processes would be carried out so that nothing would escape into the air or be flushed down the sewers. In practice this isn't easy or cheap. It may be impossible. It is impossible to operate a refinery without accumulating wastes. If they are burned, the smoke and fumes pollute the air. If pumped or hauled to an uninhabited area and dumped, they pollute the landscape. Even if they were buried, there is no assurance that circulating groundwaters wouldn't bring them to the surface sooner or later.

Environmental pollution problems continue to arise even after the marketable products from oil refineries are shipped away in pipelines, tank trucks, or seagoing tankers. The hazards of spills are ever-present. Some of the effects of petroleum fuels on the environment are inescapable. The energy in petroleum cannot be released without burning, which takes oxygen from the air and replaces it with carbon dioxide, carbon monoxide, and other exhaust gases.

Suppose, for example, that each minute sixty autos and trucks pass along a one-mile stretch of a busy street on a day when the wind is blowing at 12 miles per hour. In effect the wind replaces the air over the 1-mile stretch of street every five minutes. During a five-minute period

300 autos and trucks will have gone by and burned at least 20 gallons of gasoline. This would take 400 pounds of oxygen from the air along the street, and in exchange would have released 350 pounds of carbon dioxide and other exhaust gases plus about 160 pounds of water vapor.

Would the loss of this much oxygen and the addition of this much exhaust gas make the air significantly less fit to breathe? The answer depends on many variables, including how well the air along the street is mixed with the general atmosphere. Think of a box of air one mile long, one-fifth a mile wide, and as deep as the atmosphere is thick. Such a quantity of air contains approximately 2 billion pounds of oxygen. Obviously, the 400 pounds of oxygen used to burn 20 gallons of gasoline would not appreciably lower the amount of oxygen left in the air.

Local weather conditions sometimes keep the air from circulating or mixing. Suppose this happens along our imaginary street and the box of air in which the exhaust gases are discharged becomes only a thousand or so feet thick with winds so light that the air sits over the street for several hours. Even under these conditions it is doubtful that all the passing cars and trucks would cause a meaningful drop in the air's oxygen content.

Our noses tell us that auto- and truck-exhaust gases contain substances in addition to colorless and odorless carbon dioxide, carbon monoxide, and water vapor. Very few engines completely burn all their fuel. Furthermore, the air that goes through the combustion chambers of these engines contains nearly four times as much nitrogen as oxygen. When the explosions occur, the high temperatures and pressures are sure to form nitrogen-containing compounds which mix and may react with all the other exhaust gases. This is what we smell when the traffic is heavy and air circulation isn't adequate to carry away or scatter the exhaust gases. Under really bad conditions, the box of air above heavily traveled streets and roads turns into smog and sits like an irritating blanket over the whole countryside until changing weather conditions bring winds that sweep the smog away or updrafts allow it to scatter in the upper atmosphere.

In the United States approximately 600,000,000 pounds of petroleum products are burned daily. This calls for 2 billion pounds of oxygen. This would seem to strain the earth's oxygen supply. Actually there is almost this much oxygen in a box of air only a mile long, 1,000 feet wide, and as deep as the atmosphere is thick. Obviously there are millions of such boxes of air in the atmospheric blanket covering the country.

The problem of air pollution from burning petroleum products is primarily the addition of gases left behind in the exhaust gases. On this basis solutions to the problem would seem to await improved engineering of internal-combustion and jet engines. In addition, similar improvements are called for in the design and operation of oil-burning furnaces and all other devices which burn petroleum products.

Will the Problem of Air Pollution Solve Itself?

There is the point of view that nothing more than time will solve the problems associated with the pollution of the air by burning petroleum fuels. Depending on whose estimates you accept, the whole earth's oil supply will have been burned up in as few as thirty years or, more optimistically, as many as fifty. This suggests that we could put up with air pollution a few more years and then hope that natural processes would restore the quality of the environment.

The logic, or lack of it, in this point of view is difficult to sort out. There is only so much petroleum stored in the earth's rocks, and it is being pumped from storage areas and burned at rates known with a fair degree of accuracy. Trends permit us to estimate the rate at which the consumption of oil will increase. It is simple arithmetic to divide the total supply by the amount used each year in order to see how many years it will take to use up the supply. If we do this, we come to the conclusion that the problems associated with the pollution of the environment by the burning or spilling of petroleum and its products will be eliminated in the not-too-distant future.

It is almost impossible to imagine life in the United States without millions of automobiles, trucks, and buses speeding along highways and city streets—and large fleets of airplanes taking off and landing on daily schedules at hundreds of airports. On this basis it is difficult to accept the "logic" that time will solve the problems associated with the pollution resulting from burning petroleum products.

These problems are going to become less and less easy to ignore. They will have to be faced. Decisions will have to be made. The important thing is to be as well informed as possible, to be able to weigh the validity of one viewpoint against another, of one prediction against another. The earth's supply of petroleum has enabled the whole earth toy to spin and whirr faster during the "oil-burning era" than at any time in the history of civilization. Until a replacement for this convenient source of energy is found, there is the uneasy feeling that the whole earth toy will have to be extensively redesigned.

5 Energy from Coal

Coal is an abundant, widely distributed energy resource. Coal beds exist today where dense prehistoric swampy forests once thrived. The sinking and rising of the earth's crust caused the partially decayed plant debris from these swamps to be covered with thick layers of rock-forming muds and sands. Supposedly, when a swampy area of a sinking crustal segment became sufficiently submerged to receive the sediment eroded from increasingly emerged areas, the plants that once thrived in the swamp did not totally decay. The plant material trapped between layers of rock-forming sediments became the raw material for coal-forming chemical reactions. In time the plant tissues were converted to coal.

There are coal beds in parts of the world where the climate today supports practically no plant growth. Apparently there have been drastic changes in the conditions that determine climate. For example, tropical regions, or arctic regions covered with thick layers of ice and snow, have not always been where we find them today. There were prehistoric swamps in many parts of the world other than today's tropical and semitropical areas. This means that layers of "buried sunshine" can be found in the rock layers beneath many countries.

Next to direct sunlight, wind, and waterpower, coal is the most widely distributed energy resource. Coal is much more likely to stay put than oil or gas. Therefore, wherever coal was created hundreds of millions of years ago, coal beds are found today—unless erosion has destroyed the overlying rock layers and scattered the coal.

Onetime deeply buried coal beds can be exposed along the banks of streams or brought to the surface during mountain-building or other geological events. Primitive people may have used the black rocks that fell from exposed coal beds to build fires and, thus, "discovered" coal. Archaeological evidence suggests that coal was used in funeral pyres as many as 3,000 or 4,000 years ago. And metal workers of the Bronze Age (at least 1,000 years before the Christian Era) very probably fueled their forges with coal. Early Greek literature and the Bible provide additional evidence that coal has had a long history.

By the 1300's, so many Londoners were burning the coal that could be picked up along England's coasts that King Edward I outlawed its use. He wanted no part of a fuel that filled the air with such foul smoke and bad smells. It took several hundred years to invent ways to burn coal that allowed it to compete with wood and charcoal; but by 1660 Britishers were burning 2,000,000 tons of coal each year and nearly 8,000,000 tons by 1760.

The American Indians knew where to find coal and

Coal mining means men: A crew starts underground for its shift of work. (Courtesy of Bureau of Mines, U.S. Department of the Interior)

how to use it in making pottery. Coal was not mined in America, however, until 1760. By 1860 the annual use of coal was scarcely more than 1,000,000 tons. The impact of the Industrial Revolution hit America at about this time, and by 1900, some 270,000,000 tons of coal were being consumed each year. Its use continued to increase rapidly; by 1920 America's coal mines were turning out 650,000,000 tons of coal each year. By this time world-wide use of coal had reached 1,200 million tons.

Supply and Demand

Between 1920 and 1930, annual coal consumption dropped from 650,000,000 to 560,000,000 tons. During this ten-year period steel pipe became available, allowing the gas industry to transport gas from Western and Southern gas fields to markets throughout the United States. The production and distribution of oil expanded at the same time. In the same ten-year period the use of oil jumped from 443,000,000 to nearly 900 billion barrels annually; the use of natural gas jumped from 800 trillion to nearly 2,000 trillion cubic feet annually.

These trends in the relative popularity of oil and gas versus coal continued until, by 1970, oil was being consumed at the annual rate of 3,364 million barrels, gas at the rate of 20,600 trillion cubic feet, and coal at the rate of 560,000,000 tons.

There is an estimated 3-trillion-ton supply of coal in the United States. At the current rate of use this is enough to supply the nation for hundreds of years. The whole earth's store of coal is estimated to be well above 5 trillion tons, with annual consumption in 1970 running slightly above 3 billion tons. The world should not run out of coal for more than a thousand years.

It is puzzling that we have rushed ahead to use up nature's least abundant energy recources, gas and oil, and have tapered off our use of the most abundant energy resource. Apparently the typical user prefers to get energy in the most convenient form. The question of adequacy of the supply hasn't received much attention.

Coal and the Way People Live

Coal has played a dramatic role in changing the way people live. Coal miners, especially, felt the full impact of the Industrial Revolution—the changes that took place in the move from a basically agrarian to an industrial society. Before expanding industry created a market for coal, men and women spent most of their time growing food, weaving cloth, making bricks, processing lumber, winning metals from ore, and doing all the other kinds of hand labor needed to keep their communities going.

Learning how to dig coal and use its energy to power the world's engines was a giant step toward the Industrial Revolution. In fact, in reference to the origins of the Industrial Revolution, it might be more accurate to say that the availability of adequate supplies of coal created the industrial expansion that was characteristic of the Revolution. One person could dig enough coal to replace the muscle power of hundreds of workers and dozens of horses. In contrast to the dependence of wind or waterpower on seasonal weather conditions, coal was easily stockpiled; an industry could plan ahead for increasingly greater production.

With industrial expansion farmers tended to leave their family farms, and craftsmen left their family enterprises as people took jobs in the increasing industrialized mines and factories. People were less likely to grow their own food or make their own tools and clothes; they relied on wages to buy these things.

With these changes people in the same community came to live sharply contrasting lives. Life for those who owned the mines and factories was not the same as for the people who managed or operated them. And the more highly industrialized a community became, the greater the differences became.

There is a world of difference between tending a farm —or building a house, weaving cloth, making tools, and doing other kinds of jobs in a sunlit countryside—and being a very small part of a very large factory or going down a deep, dark hole to dig coal.

Mining Coal

All coal mines begin with an underground bed of coal. To be profitable, the coal beds should be at least three feet thick; they are rarely more than eight feet thick. They are always sandwiched between layers of shale,

Coal mining also means machinery: here is a hydraulic monitor
mounted on a Joy 8BU loading machine. (Courtesy of Bureau
of Mines, U.S. Department of the Interior)

sandstone, or other kinds of rock or clay. The coal beds may be as level as when they were originally laid down at the bottom of a swamp or lake, or movements of the earth's crust may have left them sloping or tipped. The beds may be exposed on hillsides or along stream banks, or they may be completely covered with layers of rocks hundreds of feet thick.

The character of a mine is determined pretty much by the location and condition of its coal bed. If the bed is exposed on a hillside, a drift type of mine simply follows the layer of coal deeper into the hill. In hilly areas where the beds of coal are tilted, the mine may take the form of a sloping tunnel dug through the rocks to reach the coal.

A shaft mine is used to harvest coal that lies in level beds far below the earth's surface. A hole is dug straight down to the coal beds, and then horizontal digging follows the layers of coal. All the rocks that have to be gotten out of the way to get to the coal and everything used to work the mine must be taken up and down elevators in the central shaft.

All types of mines must be provided with ventilating systems to bring fresh air to the miners and remove harmful and sometimes explosive gases that can collect in the mine. Few mines escape having water seep into them, and this water must be collected and pumped to the surface.

In modern mines specially designed machines called continuous miners break the coal loose and transfer it to shuttle cars or conveyor belts that carry it to the surface. In mines that are not so modern, the face of the coal bed

112

is attacked with huge chain saws that carve slots into the coal or with drilling-machines that bore deep holes. Explosives or cylinders of compressed air that act as explosives are then used to blast the coal loose. Loading machines then scoop and sweep the loosened coal into low-slung, electrically driven trucks or onto conveyor belts.

For people accustomed to seeing bulldozers, dump trucks, and other heavy equipment used to handle rocks aboveground, the clever engineering of mining equipment is particularly impressive. It is interesting to see how the same basic equipment can be flattened to the cramped working conditions of a mine. Even more impressive are the precautions that must be taken in the design and operation of the equipment to head off accidents.

In all types of mines precautions must be taken to keep the roof from falling in as the coal is removed. Careful decisions determine how large to make the pillars of coal that are left undisturbed to support the roof. For additional safety, as rapidly as the coal is taken from the working face of the mine, holes are drilled into the mine roof and long steel bolts are inserted through the layers of overhead rock to bind them together like sheets of plywood.

The history of the mining industry includes tragic accidents that have killed or injured many men. Cave-ins, rockfalls, fires, and explosions are constant threats, but because these threats are so obvious, adequate safety precautions are a recognized part of the whole business of

113

mining. Safety and health rules are imposed by federal and state laws. During some years, in comparison with all other industries, the mining industry shows the lowest number of accidents per million man-hours worked.

To Strip or Not to Strip

There is a type of mine that avoids many of the hazards of underground mining operations. This is the strip or surface mine. Where the beds of coal are relatively near the surface, giant power shovels, drag lines, or other earth-moving equipment move aside the overburden of rocks and soil. Once the coal beds are laid bare, smaller power shovels break the coal loose and load it onto trucks or conveyors.

The strip-mining industry supports and is supported by some of the biggest machines in the world. The largest shovels are more than 200 feet tall and weigh millions of pounds. Their dippers can take as much as 270 tons of rock at one bite and drop it more than a city block away. With these huge machines, the industry can now harvest coal by surface mining that formerly would have been mined by underground operations. In 1970, more than 40 percent of the coal burned in the United States had been strip-mined; in 1940, less than 10 percent.

Strip mining, however, solves only some of the problems of the coal industry. It protects miners from many of the serious hazards and hardships that go with drift, slope, and shaft mines. If the coal companies don't have

Two trucks are hardly a nibble for the great dipper of a giant coal stripping shovel which picks up almost 200 tons of rock and dirt in every bite. (Courtesy of Bucyrus-Erie Company)

to put the countryside back into its original condition after taking out the coal, strip mining is the cheapest way to harvest coal. But this is a very big if.

In past years mining companies were permitted to move into a countryside, shove aside or turn over the topsoil with its growth of trees, shrubs, and grass, and haul away the coal. When the coal had been stripped clean, the mining companies simply moved their expensive machines to new and not yet harvested coal beds. Stripped areas were left as ghastly scars in the countryside. Not for dozens of years would natural processes heal the wounds—not for dozens of years would the earth's soil-making processes change the jumbled mass of rocks and debris to a soil that would once again support grass, shrubs, and trees.

The damage to the countryside in abandoned strip-mining areas wasn't limited to the immediate areas where the coal had been taken. The streams suffered. Before strip mining, the streams in an area may have meandered pleasantly through the countryside. Their clear, sweet water supported interesting and sometimes valuable populations of plant and animal life. When strip mines upset the lay of the land and removed the water-retaining soil cover, the streams became murky, silt-filled sewers, so thoroughly polluted with acid mine wastes as to make them unfit for any worthwhile plants or animals. Not until these streams had flowed many miles away from the strip mines could natural processes restore their quality.

Local residents suffered. Many of these people were

accustomed to a definite way of life. Some were people whose whole lifetimes had been spent in hilly and mountainous country. The original landscape and its plants and animals played vital roles in their whole economy and philosophy.

Beds of coal beneath the surface of their beloved hills and mountains meant nothing to their way of life. Supposedly, these mountain people were content to go on living as their ancestors had lived for many generations. They felt no need for the money that could be made by converting those underground coal beds into cash value.

Much of the coal mined by stripping the hills and mountains of West Virginia, Kentucky, Illinois, Ohio, Pennsylvania, and fifteen other states is sent to distant markets. The energy resources of the coal are not used to develop the countryside where it was obtained. It is used to generate the power that builds and propels the nation's industrialized areas. On this basis the original cash value of the coal is taken from the hills and mountains of America and invested in its urban areas.

In a sense the development of the nation's urban and industrialized areas is at the expense of its hill and mountain areas. The more city people surround themselves with the comforts and conveniences of modern living, the more they destroy the environment of other people. Consequently some people are forming groups to support action to correct what they view as seriously unjust and ill-advised practices. Obviously the environmentalists can't avoid confrontations with those who own and operate the strip mines or the business and industrial interests that need abundant supplies of cheap energy resources.

Confrontations between these opposing groups have led to various kinds of action. The destruction of the countryside has been portrayed vividly and dramatically by the environmentalists who have pressured legislators and other government officials to introduce and support corrective legislation. One proposal calls for the outright prohibition of all forms of strip mining everywhere. Other proposals require the mining companies to put up enough money before they move into an unmined area to guarantee the restoration of the countryside after the coal beds have been laid bare and the coal hauled away. To provide money needed to restore areas long abandoned by defunct coal companies, it has been proposed that all future coal production be taxed.

Representatives of the coal industry portray with equal vividness their efforts to restore stripped areas. Plans for restoring the land begin before stripping shovels take their first bite of overburden. No sooner has the coal been hauled away than reclaiming experts move in. The land is leveled off, the soil is analyzed, and tree seedlings are planted or the whole area is reseeded. Helicopters or fixed-wing aircraft have been fitted with special seeding attachments whereby even the most rugged countryside can be reseeded promptly. In laboratories and research nurseries new plants are experimented with in the hope of finding species that will survive in the sometimes hostile environment of newly reclaimed land. Landscape architects work out the best locations for parks, lakes, and campsites.

Weighing more than 27,000,000 pounds, this electric stripping shovel was the world's largest land machine when it went into service in 1965. (Courtesy of National Coal Association)

In addition to hoping to solve the land-reclaiming problems, coal industrialists say, there are other reasons why strip mines will continue to be needed. Or as the National Coal Association argues in one of its brochures, "It is not realistic to expect that surface-mined coal could be replaced by coal from underground mines." To do so would call for "132 additional coal mines of 2 million tons annual capacity, a capital investment of $3.2 to $3.7 billion, three to five years before full production could be anticipated and an additional 78,000 trained underground miners."

In strip mines 80 to 90 percent of the coal in the bed can be sent to market, whereas in deep mines as much as 50 percent of the coal must be left for pillars to support the overhead. There is an estimated 45 billion tons of strippable coal in the United States, which at the current rate of strip mining would last for nearly a hundred years.

Better Ways to Use Coal

In contrast to the much smaller supplies of gas and oil in the United States, coal is reasonably abundant. For this reason scientists and engineers are looking for ways to make coal a more popular fuel. One research effort seeks to improve methods for making artificial gas and oil from coal. Coal, oil, and gas are products of the same general chemical processes. All these fuels can be traced to partially oxidized plant remains. They all are made up of carbon and its compounds.

The main differences between gas, oil, and coal are differences in the ratio of carbon to other elements, especially hydrogen. In coal there are about equal numbers of hydrogen and carbon atoms; in oil, about two hydrogens for each carbon atom; and in gas, four hydrogens for each carbon atom. As pointed out previously, there are several economically practical methods for making artificial gas from coal. Recent research has developed methods for making synthetic crude oil from coal. Pilot plants turn out about sixty barrels of oil a day from an input of one ton of coal per hour.

The sulfur content of the coal from many mines causes problems. When sulfur-contaminated coal is burned, gaseous sulfur oxides are produced, unless they are removed, the smoke adds to the problem of air pollution. This puts a special premium on coal from mines where sulfur content is low. Most such mines, however, are in the western United States rather than near the major industrial markets. For this reason research is being devoted to removing sulfur from coal before it is burned or getting rid of the sulfur oxides before they escape into the air. If this problem is solved, it will be another step toward using coal to piece out the supply of gas and oil.

One exciting idea for solving the problems that go with using energy stored in coal involves burning the coal while it is still in the original beds. The approach is to send only enough air down to the coal bed to start but not complete the oxidation of the coal. This produces flammable gases which, supposedly, will be forced from the underground coal bed and piped to the surface.

121

After the impurities in this artificial gas have been removed, the gas can be sent to market. In effect the whole undisturbed coal bed is converted to a coal-to-gas manufacturing plant. This idea promises to produce a nonpolluting fuel, with the waste materials from its manufacture left deep underground.

Ways to convert the energy in coal directly to electrical energy are the goals of additional research. The idea is to bypass the usual steps whereby coal is burned to heat water which forms steam which drives turbines which spin the generators which create electricity. Pilot plants using coal-fueled fuel cells are already operating. Magnetohydrodynamics technology, as it is called, is also being applied to efforts to convert coal energy directly to electrical energy. In this method the coal is used to produce a stream of high-pressure, high-temperature gases, which are passed through strong magnetic fields. Since the stream of gases consists of electrically charged particles, electrical energy is generated when the stream cuts the magnetic field.

Coal has played an important role in providing the energy needed to keep the stream of civilization moving. It was coal that broke people loose from a way of life in which they had only the energy from muscles and beasts of burden to do the world's work. It was coal that let people become accustomed to using more and more energy so they could live ever more comfortably. With the discovery first of oil and then of natural gas and the development of these more convenient sources of energy, people turned away from coal.

Convenience may give way to costs that are tied in with the dwindling supplies of gas and oil. Because of its abundance, coal may be brought back to play increasingly important roles in the story of man's needs and demands for energy. In the future complex, newly designed push-button-operated machines may be used to mine coal—machines with whirling steel teeth that bite and tear the coal loose from its underground beds, machines brandishing steel arms like hungry crabs, which sweep up the coal and dump it into computer-operated mine tractor-trailers. All this means that coal miners of the future will be highly skilled specialists.

6 Energy from Atoms

Atomic energy made its first great impact on the whole earth toy during the closing months of World War II. People learned of the enormous quantity of energy released through nuclear reactions when they realized the fantastically destructive effects of atomic bombs.

Actually the atomic age began many years before World War II, in the secluded laboratories of men and women who were curious about the structure of matter. The early efforts to understand what matter is and how it is put together held no hint that they would one day lead to the greatest—and most controversial—source of controllable energy known so far.

People Fear Atomic Energy

People are likely to fear what they don't understand. Something both mysterious and dangerous can be especially disturbing. What could be more frightening than to know that where atomic bombs have exploded, there are deadly invisible rays which poison the air we breathe, the water we drink, the food we eat, and even the ground we walk on? Yet countless newspaper, radio, and television stories have carried this message, often with spectacular effective dramatization.

There is another reason why many people continue to be downright afraid of atomic radiation in all forms and intensities. It, too, goes back to the circumstances under which atomic energy was announced to the general public. The tactics and strategy of war include secrecy. It was impossible to keep atomic bombs secret, but it was possible to keep them a mystery. In fact, the Manhattan Project, the code name for the enormous organization that put together the first atomic bombs, is often referred to as the best-kept secret of all time.

People naturally try to protect themselves against anything they fear. Furthermore, fear of something mysterious is easily exaggerated, so that we protect ourselves not only against real but also imagined threats to our welfare. This causes many people to be very uneasy if they learn of something coming to their community that might expose them to atomic radiation.

But here's the crunch: The development of atomic energy seems to be the most promising solution to the problems that are sure to accompany a dwindling supply of gas, oil, and coal. Nuclear reactors, however, pose some of the same radiation hazards as atomic bombs; and this means that anyone who fears atomic bombs may also fear atomic energy.

This mighty thermonuclear blast was detonated underground on July 6, 1962, at the Nevada test site. (Courtesy of NASA)

Energy Is Mysterious in All Its Forms

There is no denying that atomic energy has its mysterious aspects. But so does any form of energy. Even though the story of atomic energy has been terribly emotionalized and complicated with extreme secrecy and government security, people must think and act intelligently when faced with questions involving new developments.

This raises a sticky question. How can people explore the possibility that atomic energy might see us through the energy shortage problem and still remember that its past includes devastating destruction? Neither complacency, apathy, nor blind opposition is the answer. What seems to be called for is enough understanding of atomic and other forms of energy to enable us to trade off negative against positive factors and bring our society out the better for it.

Let's look at other situations in which energy transformations are involved. Few people really understand what happens inside the cylinders of an automobile engine when mixtures of gasoline vapor and air explode and ultimately cause the auto's wheels to turn. Many people are willing to ride airplanes even though they don't understand how jets of hot gases escaping from the combustion chambers of jet engines keep hundreds of tons of aluminum and steel flying through the air. We flip switches on all kinds of electrical gadgets that do all kinds of things without really understanding what elec-

128

trical energy is and how it is changed to heat, mechanical, or other forms of energy.

The energy that drives automobiles, airplanes, and anything else that consumes liquid or gaseous fuel is made available by reshuffling the atoms that make up complex molecules. Complex molecules of gasoline, for example, represent more energy than exists in the simpler molecules of carbon dioxide, water, and other combustion products that leave exhaust outlets. We know that the complex, energy-rich gasoline molecules were fashioned from simpler molecules millions of years ago when sunlight fell on lush prehistoric swamps—a scene far more peaceful than atomic bomb explosions. This may be part of the reason why gasoline energy is so much less fearsome than atomic energy. Actually gasoline energy has been involved in many more deaths, injuries, and costly destruction than atomic energy.

It is frightening to feel the tingle of even the slightest surge of electrical energy when we accidentally touch a live wire, and we really jump when we cause a short circuit and the sparks fly. But this doesn't keep us from continuing to flip switches on all kinds of electrically powered gadgets. Our fear of electricity doesn't bother us, nor does it matter that we don't understand the chemistry in dry cells and batteries or how whirring generators produce the electrical energy that powers our household gadgets. We may know that the electrical energy can be traced to large coils of wire spinning in magnetic fields. But the strange, invisible stickiness of magnets remains a complete mystery to most of us.

Dynamite, nitroglycerine, gunpowder, TNT—these are not terribly disturbing words to most people. The shock and noise of explosives used to blast rocks from construction sites are more fascinating than frightening. When the smoke clears and the winds scatter the hot gases, the hazards of this kind of energy change are over. If we want to, we can climb among the broken rocks and not worry about any harmful aftereffects. Although the energy-rich, unstable, and usually complex molecules of explosives must be handled carefully, the simpler molecules that remain after the explosion are harmless.

Again, any form of energy can be dangerous. Fires out of control can maim and kill. So can automobile, truck, and airplane engines. Electricity can maim and kill. So can all kinds of explosives. A single match in a baby's hands can lead to catastrophe. One drunken driver is a threat to the lives of everyone on a highway. High-voltage transmission cables can be wrecked by a storm and electrocute innocent victims. Every firecracker, pistol cartridge, stick of dynamite, every artillery shell or high-explosive bomb stored or transported can lead to amputated fingers, sightless eyes, death, and destruction.

Perhaps the dimensions of atomic energy create the deep-seated uneasiness we associate with it. We know that a single atomic bomb can equal the destructiveness of the nation's annual automobile accidents. Whereas the hottest fire that firemen ordinarily encounter may involve temperatures of a few thousand degrees, atomic bombs create temperatures of millions of degrees—temperatures hot enough to vaporize not only the fire-

Prospects for the development of nuclear fusion as a source of electric power advance in 1972 as a result of experiments using the adiabatic toroidal compressor (ATC) at the U.S. Atomic Energy Commission's Princeton, New Jersey, Plasma Physics Laboratory. (Courtesy of AEC/Princeton)

man but also every piece of fire-fighting equipment. A storm-caused short circuit in a high-voltage transmission system might shower the countryside with a few million watts of electrical energy, but a single atomic bomb can let loose more energy than is handled during an entire year by all of the nation's electrical power stations.

The Dimensions of Atomic Energy

The amount of energy that can be made available through atomic reactions brings this energy resource to the forefront when we face the possibilities of an energy shortage. We know, for example, that the fission of a weight of uranium no greater than that of a five-cent piece releases as much energy as the burning of 15 tons of coal or 3500 gallons of fuel oil. And if the day comes when the hydrogen-bomb reaction can be used to generate electricity or some other form of manageable energy, there is enough heavy hydrogen in 10 gallons of water to produce this same amount of energy. These statements sound like science fiction, but they are based on fact.

The story of the enormous amount of energy involved in nuclear reactions begins with Albert Einstein's famous equation $E = mc^2$. The equation says that energy and matter are interchangeable. It is the c^2 part of this equation that brings in the fantastically large dimensions. The c represents the speed of light. But light travels so fast that there is really no way to understand its speed by comparing it with everyday, familiar speeds. Light travels approximately 186,000 miles per second or, in the

132

metric system, 300,000 kilometers per second. A flash of light could go around the world seven and a half times in one second. Or forty times while you read the previous sentence. And although it takes days for a high-speed spacecraft to reach the moon, a flash of light on the moon could reach the earth in less than one and a third seconds.

To add to the difficulty of comprehending the dimensions involved when matter is converted to energy, remember that the speed of light is squared in the $E = mc^2$ equation.

We understand a pound of butter or a pint of water. Even less precise terms such as "a dollar's worth of gasoline" or a "yard of cloth" are reasonably clear. Not so for the units that measure energy, unless you have studied or worked with physics or some branch of the energy sciences. To become familiar with the most commonly used units for measuring energy, think about three very familiar energy sources: food, gasoline, and electricity. We know how much energy it takes to see us through a day and how much food we must eat to provide this energy. On a day-by-day basis, people consume enough food to yield between 1,000,000 and 3,000,000 calories (if we define a calorie as the amount of heat needed to raise the temperature of one gram of water one Centigrade degree). Food energy, however, is more commonly measured in kilocalories (that is, the amount of heat needed to raise the temperature of a kilogram of water 1 Centigrade degree). It is this unit that is used when a person speaks of going on a 1,500-calorie diet.

133

We know that a gallon of gasoline contains enough energy that when it is burned in the cylinders of an automobile engine, the auto can be driven about 15 miles. This represents 30,000,000 calories.

The monthly electric bill on the author's desk at the moment is for 1,000 "kilowatt-hours used." At the rate of approximately 33 kilowatt-hours per day, this is equivalent to 30,000,000 calories.

How Much Energy per Person per Day?

Counting these three everyday uses of energy—food, gasoline, and electricity—a person's daily energy requirements might add to 62,000,000 calories. But food, gasoline, and electricity account for less than one-fourth of an average person's share of the energy used to keep our nation going. Three times this much energy is involved in each person's share of the energy required to operate public transportation systems; to heat and maintain public buildings; to run the factories and businesses that make and distribute the countless materials and gadgets we use every day; to equip and maintain educational and research institutions; to build, equip, and operate hospitals, churches, radio and TV stations; to write, print, and distribute books, newspapers, and magazines; and to supply all the other energy-demanding activities that add up to an ongoing society.

When we include all the demands our society makes on the whole earth's energy supply, each person's daily share adds up to 240,000,000 calories. This may seem an overwhelmingly large number, but its meaning is sufficiently familiar to help us understand the very large amounts of energy involved in atomic reactions. How much atomic fuel—uranium 235, for example—would it take to yield 240,000,000 calories of energy? Assuming nuclear reactors are about as efficient as other energy-conversion devices, the answer is approximately one-fiftieth of a gram of uranium. On this basis, a chunk of uranium 235 that weighs as much as a five-cent piece could be the fuel that is needed to provide an average person's energy needs for 250 days.

How does this fit into the $E = mc^2$ equation? According to this equation, if one gram of any substance is converted to energy, 20 trillion calories of energy are produced. But when uranium is used as the fuel in a nuclear reactor, its atoms are split into smaller atoms rather than totally converted to energy. However, when the fission occurs, approximately one-thousandth of a gram of the uranium in a gram of the fuel is changed to energy. The other 999-thousandths of the original gram of uranium now exists in the form of atoms of other elements. These are the so-called fission byproducts.

Other Units for Measuring Energy

If we use pounds instead of grams and Fahrenheit rather than Centigrade degrees, energy can be measured in units called British thermal units, or BTU's. One BTU is the amount of heat needed to raise the temperature of one pound of water one Fahrenheit degree. A pound is 454 times larger than a gram, but a Fahrenheit degree is only five-ninths as large a temperature change as a Centigrade degree. Thus, one BTU represents about 252 times as much energy as one calorie. The heating and cooling capacities of furnaces and air conditioners and the fuel value of gasoline are usually expressed in BTU's—but we don't hear dieters counting their BTU's.

There are other units that feature the effect of energy in getting things to move. These units are ergs, joules, and foot-pounds. One erg is the amount of energy needed to get a mass of one gram moving at a speed of one centimeter per second, where the only resistance is the inertia of the one-gram mass. If a kilogram mass is involved, and the acceleration increased to a meter per second per second, the amount of energy is one joule. Changing to the English system, one foot-pound of energy is the amount needed to give a one-pound mass an acceleration of one foot per second per second.

The quantity in any amount of energy can be expressed in any of these units. However, calories and BTU's are most commonly used when the energy is involved primarily in the form of heat; the other units are used when the energy is involved in moving things or in getting work done. All the units are interchangeable. For example, a person could talk about counting ergs just as he talks about counting calories. There would be problems, however, since one calorie is equivalent to 41,800,-000 ergs.

Atomic Fusion as a Source of Energy 15 [3]

Splitting or fissioning atoms is not the only way to release atomic energy. There is also atomic fusion. In this case simple low-mass atoms are brought together, or fused, to produce larger, more compact atoms. As with atomic fission (and in keeping with the $E = mc^2$ equation), if the mass of the fused atom is less than the sum of the masses of the smaller atoms, energy will be released.

Hydrogen is the element most often associated with atomic fusion. Its atoms are the simplest that exist. The predominant kind of hydrogen atom consists of a single proton-electron pair. A small fraction of the atoms in any sample of hydrogen have one or two neutrons combined with the proton. These isotopes, deuterium and tritium, as the names suggest, have twice or three times the mass of the predominant type of hydrogen atoms.

If these heavier hydrogen isotopes are brought together under the required conditions, they can fuse to form helium atoms. This is the nuclear reaction of the hydrogen bomb. It is an energy source that stretches our imagination even farther than do the enormous quantities of energy liberated by nuclear-fission reactions. For example, someone has calculated that if the deuterium in one quart of seawater could be extracted and used as fuel in a nuclear-fusion reactor, it could produce 8 trillion joules of energy. This is enough energy to meet the needs of an average person in the United States for more than twenty years.

137

Atomic fission and fusion represent a tremendous energy resource. To make the energy in this resource available, however, calls for the solution of many exceedingly complex problems. In order for nuclear energy to replace the world's dwindling supplies of gas, oil, and coal, adequate quantities of nuclear fuels must be located, nuclear-reactor power stations must be built, and ways found to distribute the energy they produce to meet the needs of people in all kinds of situations.

The Earth's Nuclear Energy Resources

The most important fissionable fuels are uranium and thorium. How much of these elements is there in the earth's rocks? And how much is in ores rich enough in uranium or thorium to allow the metals to be extracted at reasonable cost? The heavy isotopes of hydrogen, deuterium and tritium, provide the fuel for fusion nuclear reactions, and as previously pointed out, there is an almost unlimited amount of these fuels in the world's oceans.

The fissionable fuels, especially uranium, exist in many different kinds of rocks in many different parts of the world. The problem is that uranium and thorium are present in very low concentrations in all except a few rock formations. To estimate the quantity of this energy resource, therefore, we must take into account the difficulties encountered when the ores of uranium and thorium are to be harvested.

Uranium is currently obtained from mines more than

a mile deep, and ores are considered commercially profitable if they contain approximately 6 pounds of uranium per ton of ore. It is assumed that men will continue to be willing to work in mines more than a mile deep and that engineers will continue to improve the procedures for extracting uranium from ore. Under these assumptions, the amount of recoverable uranium ore in the United States that contains more than 0.17 percent uranium is estimated to be between 800,000 and 2,500,000 tons.

There is a much more abundant supply of thorium, but it occurs in even lower concentrations than uranium. There is an outcropping of granite in New Hampshire, for example, that contains approximately 0.005 percent thorium. The outcrop extends over 300 square miles, and this single granite formation contains millions of tons of thorium. But for every ton of thorium recovered from this ore, at least 20,000 tons of granite would have to be handled, and this means not only cutting the granite from its formations but taking it apart atom by atom to extract the thorium.

Whether the thorium in these granite formations can be a practical energy resource depends on many circumstances. It is known that if the thorium in one ton of 0.005 percent thorium ore is extracted and used to fuel a nuclear reactor, it could produce as much energy as 150 tons of coal or 650 barrels of petroleum. Knowing the cost of coal and petroleum, a businessman can calculate the market value of the thorium he might expect to obtain from processing the thorium ore. It is very difficult, however, to estimate how much it will cost to mine the granite and extract the thorium. And there are still many problems to be solved before the United States will be ready to go ahead with the installation of nuclear reactors as a principal source of the nation's energy. Until these problems are solved, the extraction of thorium from granite will be a risky business proposition.

The United States Atomic Energy Commission estimates that 243,000 tons of uranium fuel can be taken from the rocks in the United States at a cost of less than $10 per pound. The worldwide reserves of equally profitable uranium ores amount to an estimated 840,000 tons. Actually the cost of any fuel source can't be accurately predicted very far into the future. The adequacy of the supply versus the urgency of the demand can change rapidly in a society that is quick to take on new energy-consuming gadgets and practices. Furthermore, the cost of nuclear fuels threatens to climb when the richest and most easily processed ores are used up. But when other energy resources—especially gas, oil, and coal—become scarcer the market value of the uranium will increase.

Manson Benedict, writing in the September, 1971, issue of the *Bulletin of the Atomic Scientists*, estimates that if uranium replaces other fuels for generating the nation's electricity, by the year 2000 the price of uranium will have risen from the current $8 to more than $15 per pound. He also points out that to generate all the electricity the United States will need between now and the year 2000 would burn up enough uranium to cause a serious shortage of readily available uranium ores.

140

Heading Off a Shortage of Nuclear Fuels

Realizing that nothing is to be gained by replacing a shortage of fossil fuels with a shortage of nuclear fuels, some energy futurists warn against burning up the nation's supply of uranium in the ordinary nuclear fission type of reactor. They say that the uranium taken from rocks should be stockpiled and that no more ores should be harvested until engineering technologists have solved the problems involved with the new breeder type of nuclear reactor.

Seemingly contrary to the usual concept of energy conservation, the breeder reactor creates more nuclear fuel than it consumes. The explanation involves straightforward science.

When uranium 235 is used as the fuel in an ordinary nuclear-fission reactor, each uranium 235 atom that splits and releases energy is reduced to at least two atoms of lower-atomic-weight elements. These fission byproducts are no longer useful as fuels for power-generating reactors, at least on a large scale. As has been pointed out, only one-thousandth of each gram of the original uranium fuel is converted to energy, and 999-thousandths become fission byproducts. The breeder type of nuclear reactor finds a very effective use for some of the particles that appear among these fission byproducts.

In any sample of uranium taken from its ores, there is only one uranium 235 atom for every 140 uranium 238 atoms. There is an important difference between these two uranium isotopes. Uranium 238, with three additional neutrons in the nuclei of its atoms, cannot be used to fuel a nuclear-fission reactor. In fact, since these isotopes have almost identical chemical properties, learning how to separate the fissionable uranium 235 from the other isotopes that make up uranium was the breakthrough that made nuclear energy available.

The secret of the breeder type of nuclear reactor lies in using the scarce uranium 235 to produce, or breed, other kinds of nuclear fuels. For example, useless, nonfissionable uranium 238 can be converted to fissionable plutonium-239, or reasonably abundant thorium can be converted to fissionable uranium-233.

In theory the procedure is remarkably simple. Here's how it is done. Build an ordinary nuclear-fission reactor. Fuel it with the usual uranium 235. But pack the reaction chamber with uranium 238 or thorium 232. When the uranium 235 atoms undergo fission, in addition to the smaller atom-fission byproducts produced, high-speed neutrons are hurled from the nuclei of the uranium 235 atoms. Some of these neutron bullets find as their targets the nuclei of nearby uranium 239 or thorium 232 atoms.

Before the uranium 238 target atoms were hit by bombarding neutrons, their nuclei consisted of 92 protons and 146 neutrons—a reasonably stable, nonfissionable arrangement of protons and neutrons. But when one more neutron crashes into the nucleus of this kind of atom, things happen. Two neutrons lose beta particles. These beta particles are best described as bundles of negative electrical energy. In theory each beta particle in a neutron is balanced against an equal amount of positive electrical energy. The loss of a beta particle by a neutron—that is, the loss of negative electrical energy—changes its character from that of a neutron to that of a proton. χ

The number of protons in the nucleus of an atom sets the characteristics and name of the atom. Regardless of the kind of isotope involved, each uranium atom, for example, has 92 protons in its nucleus. Change this number to 94 and we have a new kind of atom. It is an atom of plutonium. More important from an energy-resource point of view, plutonium 239 atoms are fissionable and can be used to fuel a nuclear-fission reactor.

The situation involving thorium is similar. A thorium 232 target nucleus, before bombardment, consists of 90 protons and 142 neutrons. Following impact by a bombarding neutron, two neutrons lose beta particles and become protons. We now have an atom of fissionable uranium 233.

In practice when plutonium 239 or uranium 233 are used to fuel a reactor, they end up producing more new fuel than they consumed. For each atom that splits when these kinds of atoms undergo fission, two or three neutrons are hurled toward nearby atoms. And each bullet neutron stands a chance of finding a target if the reactor vessel is packed with uranium 238 or thorium 232.

While a breeder type of reactor is operating, not only does it produce energy, but it also creates new fuel. This fuel can be used to keep the reactor running, or it can be shipped to other locations. Not until the earth's store of uranium and thorium has been used up will this energy resource be totally exhausted. In Manson Benedict's opinion, this could be far, far into the future. He estimates, for example, that there is enough uranium in the United States to provide the fuel needed to generate electricity at the rate it is being used today for 64,000 years, assuming that all the available uranium is set aside to use as fuel in breeder reactors.

An experimental breeder reactor operated in the United States between 1951 and 1963. Two pilot-plant breeder-reactor electrical-power-generating plants have been operating since 1965. Similar setups have operated in England since 1963, in France since 1967, and in the USSR since 1970. Additional installations are under way in these and other countries.

It appears that at least a partial solution to the problems of an impending worldwide energy shortage has already been achieved. There is the suggestion that the earth's supply of uranium and thorium can be used to meet the world's needs for electrical energy. This would allow the remaining supplies of gas, oil, and coal to be reserved for other energy needs. These energy resources would then last for many more years. However, these ideas arouse controversy for several reasons. For one, many people are downright afraid of nuclear energy in any form or in any amount.

Are Nuclear Reactors a Safe and Sane Source of Energy?

How safe are nuclear reactors, especially the breeder type? The answer depends on whom you ask. Some experts even dodge the question. Many people prefer their own interpretations to the experts' answers, interpretations likely to strengthen previously adopted points of view, prejudices, beliefs, and opinions.

One reaction is to dodge the question by asking: How safe is any large-scale source of energy? Don't steam boilers blow up? Aren't people killed or injured every day in accidents connected with other kinds of energy-conversion systems? Aren't there risks and hazards in all efforts to produce and distribute energy? Wouldn't it be better to go ahead with the development of nuclear energy than to run the risk of exhausting our relatively less hazardous energy resources?

Actually the questions raised in these arguments call for intelligent examination of many facts. For example, the more people know about the theory and design of nuclear reactors, the better they can weigh one hazard against another or the advantages against the risks.

Basic to the theory of nuclear-fission reactors is bringing together under controllable circumstances a chunk of uranium 235 or other fissionable fuel large enough to cause a chain reaction. In any quantity of uranium 235, a known fraction of its atoms are disintegrating each moment. Each disintegration releases energy and creates

145

subatomic fragments, including neutrons. Each neutron hurled from the nucleus of a disintegrating atom stands a chance of crashing into the nucleus of a neighboring uranium 235 atom. Whether a neutron finds another uranium atom depends on the size of the chunk of uranium 235. When the total mass of uranium 235 is less than critical, the neutrons hurled from the nuclei of the disintegrating atoms don't fission other uranium 235 atoms and no chain reaction occurs.

When the chunk of uranium 235 is large enough to cause a chain reaction, the combination of accumulating energy with the ever-increasing barrage of high-energy neutrons let loose results in an almost instantaneous fissioning of the chunk of nuclear fuel—and the accompanying release of a tremendous quantity of energy.

Under controlled conditions less-than-critical-mass quantities of the nuclear fuel are packed into long, carefully sealed metal tubes. The reactor chamber is built in such a way that these fuel units can be moved toward or away from the center under closely controlled circumstances. This permits enough fuel to be brought together to form the critical mass. When too heavy a barrage of high-energy neutrons develops, the fuel rods are withdrawn and the neutrons spin off their energy without starting a chain reaction. This procedure also keeps under control the amount of energy being released.

Nuclear-fission reactors are designed and operated to yield usable energy. To collect the energy from the area where the actual fissioning of the fuel occurs, one system pumps water under high pressure through the reaction

146

chamber. The water enters the chamber at a temperature of over 500°F. and leaves at temperatures above 600°F. This pressurized hot water then heats other water before being recirculated through the reaction chamber. The heated water, in turn, produces steam to spin the turbines that drive electrical generators.

So long as the atomic fuel in the separate fuel rods is kept from crunching together sufficiently to produce critical mass, there is no risk of the reactor's becoming an uncontrolled "atomic bomb." On this basis, nuclear reactors are judged safe for making large quantities of energy available. Furthermore, fail-safe engineering arrangements ensure that a critical mass of fuel cannot be crunched together accidentally. In a fail-safe situation, when an accident causes a safety device to fail to work properly, the failure itself creates conditions which automatically restore safe conditions. For example, if something happened to the control mechanisms and the fuel rods were accidentally driven into the reaction chamber too rapidly, the resulting increase in the amount of fuel consumed would automatically force the fuel rods away from the reaction chamber. A critical mass of fuel would thereby no longer exist, and the reactor would shut down automatically.

Or suppose the circulating pressurized water became abnormally hot. The dangerously high temperature would be tied with control mechanisms and the fuel rods would be withdrawn. Whatever accident caused the hot water to fail to circulate properly would also cause the control rods to be restored to safe positions.

Nuclear reactors are not without hazards. All situations in which large amounts of energy are involved are hazardous. Any quantity of energy out of control can cause damage. Accidental shotgun blasts maim and kill. So do short circuits in household wiring or electrical-appliance cords. Auto and truck engines allowed to get out of control continue to release energy until they destroy themselves and maim or kill their drivers.

In today's society few energy-transforming systems are equipped with fail-safe precautions to ensure that accidents automatically bring out-of-control energy back under control. And engineers cannot be sure that they have anticipated all possibilities of failure and provided the required fail-safe precautions for any energy-transforming system, including nuclear-fission reactors.

Radioactivity: The Invisible Hazard

One hazard associated with nuclear reactors continues to worry many people. Whenever atoms undergo rearrangement of the particles that make up their nuclei, there is always the possibility that atomic fragments or bits of energy will be hurled from the unstable atoms. This is radioactivity, and there is certainly a great deal of radioactivity during the operation of a nuclear reactor.

But engineers can build into the design of nuclear reactors the safety features necessary to protect the operators from this radiation hazard. Fission byproducts or waste materials create a much more difficult problem. When a uranium 235, plutonium 239, or other atomic-fuel atom is fissioned, the two newly created atoms are rarely stable. There are only certain combinations of protons, neutrons, and other particles which build atoms that produce stable, nonradioactive atoms. When atomic particles are brought together in other than these certain ratios or combinations, the particles rearrange, and if necessary, excess particles are hurled from the nucleus to achieve a stable condition. This is usually the situation with fission byproducts.

For example, if a uranium 235 atom splits into approximately equal halves, each half receives 46 protons. If the 144 neutrons are equally distributed, 72 go into each new atom. Atoms with 46 protons form the element palladium. But a stable palladium atom contains only about 60 neutrons. On this basis, the newly created fission-byproduct palladium atoms would be highly radioactive.

For a second example, a uranium 235 atom might fission in such a way that 45 protons would go to one fragment and 47 to the other. This yields one atom each of rhenium and silver. There is a total of 119 neutrons in a stable atom of each of these elements. If the bombarding neutron that caused the uranium 235 atom to fission became trapped in the target nucleus, then there would be 25 more neutrons than needed for the atoms of rhenium and silver. Consequently the newly created fission byproducts would be radioactive, and they would eject the leftover bits and pieces until stability was reached.

Some fission byproducts become stable and thereby lose their radioactivity very quickly—that is, in fractions of a second. Other fission byproducts do not complete their nuclear rearrangements for many years. This is why it is such a difficult problem to get rid of the wastes from a nuclear reactor.

To add to the hazards associated with radioactive wastes from nuclear reactors, most chemical processes don't discriminate between ordinary, stable atoms of an element and the radioactive forms of the same element. This is especially significant when we are dealing with the chemical processes involved in living systems, particularly our own bodies.

To illustrate, suppose that radioactive calcium has accidentally been scattered over the countryside. The grass that grows in the area doesn't discriminate between this radioactive calcium and the nonradioactive calcium ordinarily present in the soil. In time the radioactive calcium is absorbed into the tissues of the grass. When cows eat this grass, both kinds of calcium are handled in the same way throughout the cow's body processes. When a person drinks some of the cow's milk, the person receives both kinds of calcium atoms, radioactive as well as nonradioactive, stable calcium atoms.

Every radioactive calcium atom built into the tissues of a person's body can be a health threat. Suppose the atom is built into a complex molecule that plays a vital role in one or another life process. Before that molecule gets the chance to function normally, the radioactive calcium atom undergoes a rearrangement of its nuclear particles. Specifically, suppose it loses a beta particle. This in effect adds another proton to the atom's nucleus, chang-

150

ing it to an atom of scandium, which has chemical properties that are quite different from those of calcium. Consequently the new atom no longer fits into the original complex molecule that was built around the one-time calcium atom. This is enough to destroy the whole molecule or at least its proper role in body processes.

Radioactive atoms taken up in the tissues of a person's body pose a second threat. The particles hurled from disintegrating atoms are effective bullets for taking apart whatever molecules they hit. A beta particle, for example, whizzing through a nearby molecule is very similar to the electrons which, because they are shared between the atoms in the molecule, hold it together. This means that the beta particle is in effect a threat to the glue that holds the molecule together. Broken molecules, in turn, threaten normal physiological processes.

Suppose a radioactive calcium atom is built into bone tissues near a man's testes. If the atom undergoes a rearrangement of its nuclear particles and in the process ejects a beta particle, the beta particle might spin off its energy before it hits any vital molecule. On the other hand, the beta particle might penetrate the tissues of the testes and hit a molecule that is part of a gene in one of the chromosomes of a sperm cell. The atoms making up this molecule could thereby be rearranged. The molecule would take on different properties, and whatever trait was controlled by the gene could be changed—that is, undergo a mutation.

The sperm cell containing the mutated gene may happen to be the one that fertilizes an egg cell. If so, there is the chance that the characteristics of the resulting off-spring will have been influenced by the original radio-activity of the calcium atom. This reasoning may have caused many people to be afraid of all kinds or amounts of radioactivity.

People who understand the whole story of gene mutations and their threat to the characteristics of not-yet-born offspring, however, tend to have different attitudes. These people take into account the statistical probabilities involved. They know what the chances are, for example, that a single beta particle will find a physiologically significant target. Or what the chances are that a specific sperm cell will be the one that fertilizes the egg that becomes an embryo.

Furthermore, these people know that all body tissues are constantly exposed to radioactivity from natural sources. Many of the elements essential to body processes include radioactive forms, which become caught up in body tissues. Any ordinary environment includes sources of radioactivity from unstable atoms that are built into rocks and soil. And cosmic rays continually expose body tissues to a kind of bombardment that can create effects very similar to radioactivity.

The fact remains, however, that all radioactive wastes from nuclear-fission reactors must be disposed of properly. This means that they must be locked up forever so they can never get into the usual flow of materials through the biochemical world. If not forever, for at least the number of years required for the longest half-life elements to lose virtually all their radioactivity.

To find a safe place to store radioactive wastes poses a difficult problem. Abandoned salt mines offer one solution. It is obvious that water doesn't circulate through these mines; otherwise the salt would have been washed away. The storage containers could be protected against all the usual causes of rust and corrosion. It is also assumed that people could be kept from accidentally stumbling into the storage area or tampering with the stored materials.

Another proposal would seal the wastes in thick concrete pillboxes and set them aside in remote storage areas. The areas would be marked with appropriate signs, and the wastes would supposedly stay locked up until the cement crumbled away, something that wouldn't happen for hundreds of years.

There are other ideas. Someone has suggested loading the wastes into rockets and launching them into outer space. Or sealing them in cement barrels and dropping them into the deepest part of the ocean. Or hauling them to the remote Arctic. In general, however, no proposed solution to the storage problem convinces many people that there won't be serious radioactive hazards if the nation goes in for large-scale nuclear-fission reactors as a way to head off the energy shortage.

What About Atomic Fusion as a Controllable Source of Energy?

To many people, the promise of obtaining energy from nuclear-fusion reactions is extremely challenging.

And for several reasons. The hope of finding almost unlimited quantities of energy is the greatest one. There is also the fact that fusion reactions need not produce radioactive byproducts or waste materials. ⁄

If the heavy isotopes of hydrogen are used as fuel in a nuclear-fusion reaction, the only product left in the reaction chamber is helium. This helium is nonradioactive. For each 4.03 grams of deuterium converted to helium, for example, only 4.004 grams of helium are produced. Of the original weight of deuterium, 0.026 grams are converted to approximately 500,000,000 calories of energy. This is the basis for such statements as: If the deuterium in one quart of seawater could be used to fuel a fission reactor, enough energy would be produced to provide the total energy needs of an average person in the United States for more than a month. And there would be no radioactive wastes to get rid of.

But there are extremely difficult technological problems to solve before fusion energy can be harvested and distributed throughout the usual energy-transfer systems. Before deuterium atoms can be made to fuse and release energy, the fuel must be heated to temperatures above 1,000,000 degrees. Heated this hot, the deuterium atoms that crash into each other will stick together, or fuse, and become helium atoms. The conversion of mass to energy in turn provides heat that keeps the reaction going until all the deuterium fuel is used. Consequently once the reaction is started, temperatures far above 1,000,000 degrees develop. But no known substances can stand temperatures this high without being immediately boiled away into flaming gases.

154

One very difficult problem in connection with nuclear-fusion technology is to construct some kind of chamber to contain the reaction. In hydrogen bombs there is no need to worry about a container: The reaction is a one-time event, and the intent is not only to destroy the reaction chamber but also to smash to smithereens everything else in the neighborhood. This is not the purpose of building fusion reactors for the release of energy—energy to be controlled and made available day after day, year after year, and—with luck—century after century.

It isn't easy to build a container for something so explosively hot that it converts all known materials into flaming gases. One idea takes advantage of the properties of these gases, better known as plasma, to build their own container. The 1,000,000-degree temperatures which build up during the fusion reaction strip electrons from all the hydrogen fuel, as well as from the newly forming helium atoms. This means that the protons remaining in the nuclei of the atoms in the fusion-reaction area make the resulting plasma electrically unbalanced—that is, electrically charged. In effect the swirling plasma has the properties of an electrical current. One of these properties is the creation of a magnetic field along the path of the current, with the direction and strength of the magnetic field controlled by the quantity and direction of the plasma.

Magnetic forces can be made to oppose each other. Anyone who has played with magnets knows that like magnetic poles oppose each other. One type of container for the plasma involved in the hydrogen-fusion reaction takes advantage of this concept. Simultaneously with the

155

buildup of temperatures adequate to start the fusion reaction, magnetic fields are built up in the area of the
reaction. Whenever the plasma seems to be escaping, its
motion generates magnetic forces which oppose those of
the surrounding magnetic-force bottle. Under these circumstances all the plasma ever actually touches is the
surrounding magnetic field. Furthermore, unexpected
forces which might cause the plasma to move from the
magnetic bottle generate additional opposition between
the plasma and its self-generated container.

One promising solution to the containment problem
creates the surrounding magnetic field in the shape of a
baseball seam. The seam surrounds the core of the baseball in such a way that all points on the seam are equidis-
tant from the center. This means that a bit of something
at the center of the baseball cannot escape without getting closer to or farther from other points on the seam. If
the seam were elastic and tied to the center, any stress on
the seam at any point would force the center to stay put.

In practice the properties of plasma are far less pre-
dictable than is the core of a baseball. The plasma can
react with the surrounding magnetic field in mysterious
ways, and the invisible walls of the bottle can break
down. Sometimes the plasma creates localized or whirl-
pool magnetic fields that open up escape routes by attraction rather than opposition to the surrounding magnetic
forces. All in all, it takes time to do the research to
understand the behavior of matter under the fantastically high temperatures that develop when nuclear
fusion occurs.

It may be that the containment problem can be by-passed. One idea in this direction would feed atomic fuel into the reaction in such tiny quantities that the energy is released in bursts small enough to be mangeable—that is, to be converted to electrical, steam, or some other conventional form of energy.

The story of man's efforts to understand and eventually use the enormous quantities of energy that can be released through nuclear reactions provides a beautiful illustration of how scientific and technological research works. The whole episode began out of sheer curiosity. It will culminate in providing enough energy to allow man to develop a civilization that is limited only by the creativity, inventiveness, and hopes of the world's people. The episode has brought out the pessimism of those who were afraid the nuclear energy project would never outlive the destruction that so dramatically marked the opening of the atomic age. To optimists, there is no escaping the challenge in such statements as: If only one-hundredth of the deuterium in the oceans could be extracted for use as fuel in nuclear reactors, the energy released would amount to 500,000 times the energy stored in the whole earth's supply of gas, oil, and coal.

7 Other Forms of Energy

The winds, rivers and waterfalls, the oceans' tides, and volcanoes, hot springs, and geysers create energy resources available to man ever since the beginning of civilization. They will always be available. They cannot be used up. With the possible exception of geothermal energy, they cannot be depleted. The winds will always blow. Rivers will always flow. The tides will always rise and fall. It is difficult to believe that volcanoes or hot springs and geysers will ever dissipate the thousand-degree temperatures in the rocks deep in the earth's crust.

Although they have always been available, man has not always made the greatest possible use of these natural energy resources. Wind and waterpower, for example, were popular energy sources early in the history of man. In many European countries and especially in the early days of the United States, thousands of towns' and villages had mills and factories driven with waterpower. And windmills were a characteristic feature of the landscape of many countries.

The popularity of these natural energy resources is undoubtedly tied to the availability of competing energy sources, especially fossil fuels. Before gas, oil, and coal were discovered and the necessary technology developed for their distribution, waterwheels and windmills were popular. In addition to wide distribution, these energy sources can be exploited with relatively simple, easily engineered devices.

Geothermal and particularly tidal energy resources have not yet been much more than experimentally developed. Not only are these energy sources less widely distributed, their development seems to call for more complex, expensive, and difficult-to-design installations.

The impending shortage of the fossil fuels—gas, oil, and eventually coal—revives interest in the earth's minor energy resources. The popularity of wind and waterpower may be restored. Or the rising cost of fossil-fuel energy will increase the financial attractiveness of investing in the research that must be completed before tidal energy resources can be fully exploited. And some energy experts enthusiastically support the idea that geothermal energy resources deserve the investments that must be made before this resource can be exploited.

The Source of Wind Power's Energy

The energy of the winds can be traced to interactions between the sun's energy falling on the earth and the earth's gravitational forces. So long as the earth rotates and revolves, different areas of its surface receive unequal amounts of solar energy during different times of the day or seasons of the year. This means that the earth's

160

atmosphere and its waters are heated unequally. When air masses are heated or cooled at different rates, they expand or contract unequally. With unequal changes in volume, the earth's air masses take on unequal densities; warmer masses become less dense than the cooler air masses. Gravitational forces cause the more dense masses to settle, the less dense to rise. Winds are the earth's air masses in motion.

To illustrate this complex set of interacting forces and circumstances, think of a northern lake still covered with the winter's ice and snow. When spring comes, the increased input of the sun's energy will melt the snow and warm the surrounding countryside more rapidly than it warms the surface of the lake. In turn, the air over the land areas will become warmer than the air over the lake. The warm air will rise as it is displaced by the colder, denser air over the lake.

A somewhat different situation may develop later in the spring or during the summer. During daylight the land areas may be warmer than the lake's surface. When the sun sets, however, the land areas—and the air over them—cool more rapidly than the water. In turn, this cooler, more dense air will move toward the lake and displace the warmer, less dense air.

In either case, whether it is a land breeze or a sea breeze, the movement of the air results from an interaction between the variations in the input of the sun's energy and the earth's gravitational forces. Involved in this interaction is the additional fact that the same input of the sun's energy raises the temperature of water less,

161

pound for pound, than it raises the temperature of rocks and soil or any other material.

Using the Energy of the Winds

Tornadoes, tropical storms, and other types of destructive windstorms provide evidence that the earth's winds represent a tremendous quantity of energy. In more peaceful settings the gentle tug of the wind on a high-flying kite or the quiet urging of the wind-filled sails of a skimming sailboat reveal the energy stored in moving air masses.

Between the fifteenth and nineteenth century much engineering skill was invested in designing machines that would effectively capture the energy of the wind. In time windmills became common landmarks in many countries, especially Holland, Belgium, and the newly settled, wind-swept plains and prairies of the western United States.

The European type of windmill usually had a small number of very large canvas-covered vanes or sails. In America as many as twenty or more small metal vanes were used to catch the energy of the passing wind. In both types, cleverly engineered devices kept the vanes head on into the wind. Other devices adjusted the pitch of the sails or vanes to fit the speed of the wind and the size of the load the spinning wheels were made to carry. Well-designed gear systems protected the windmill during excessive winds and transferred the energy of the whirling vanes to where it was needed to pump water, grind grain, or do other useful work.

When windmills were a popular solution to man's energy needs the engineering literature of the day had much to say about the science and technology of windmills. It is known, for example, that the maximum power that can be gotten from a windmill increases directly as the cube of the wind speed. (On this basis a ten-mile-per-hour wind will produce eight times as much power as a five-mile-per-hour wind.) The number of turns a windmill will make in a given time varies inversely as its diameter. There is no gain in efficiency by having the sail area greater than seven-eights of the area swept by the sails.

The science and technology of wind and windmills received much attention when wind power was a popular energy resource. With the coming of more convenient energy resources—coal, oil, and gas, for example—scientists and engineers abandoned their interest in wind power. Has the potential of wind power been explored fully? The winds sweep nearly all of the earth's surface. There are millions of places where windmills could be used to provide at least small quantities of energy. And if windmills were good financial investments in earlier days, might the day not come when windmills will return to the landscape of Holland, Belgium, Kansas? Or become new signs that newly developing nations are working out their energy-supply problems in spite of the whole earth shortage of fossil fuels?

History proves that a family can build a windmill that will harness enough of the wind's energy to meet many of its energy needs. Water can be pumped into elevated storage tanks and the windmills can be made to drive generators that charge storage batteries to tide the family over windless days. It is problematical, however, that wind power could ever operate central power stations to meet the energy needs of towns or cities. Not that there isn't energy in the earth's winds. The problem is to let them blow when they will and still produce a steady source of energy for the "customers."

One proposed solution to this problem is simple. Use windmills to drive electrical generators. Use the electricity generated to separate water into its two components, hydrogen and oxygen. When the winds blow, stockpile the hydrogen and oxygen. On calm days fall back on the accumulated supply. Electricity can be generated by using cells fueled with hydrogen and oxygen. Furthermore, the experience being gained by aerospace engineers with hydrogen-fueled engines may improve the efficiency of the processes whereby the chemical energy in hydrogen and oxygen can be made available.

The Rise and Fall of Waterpower

Before the days of coal-fired steam engines or petroleum-fueled internal-combustion and jet engines, waterpower was a popular solution to man's energy needs. Whenever possible, towns and villages were settled along streams, and "down by the old millstream" was a meaningful phrase.

Quite early in the history of civilization people caught onto the fact that anywhere that water flows or falls is an energy resource. A person needs only to step into a running stream to feel the energy of running water. To be bashed and tossed about by the water that cascades over a waterfall or whirls through rapids convinces us that large quantities of energy are involved in rapidly moving water.

The energy of flowing water can be traced to the water cycle—that is, to the circulation of water through the atmosphere and hydrosphere. It begins with the sun evaporating water from the oceans, lakes, streams, and from the moist earth. Water vapor is less dense than air and is consequently displaced upward by gravitational forces. The vapor-filled air is cooled, however, as it rises, and this cooling lowers its capacity to hold water vapor. If the air becomes cool enough, the water vapor condenses to form fog droplets, rain, or some other form of precipitation. The precipitation falls, and the water continues to move in response to gravitational forces until it approaches as closely as possible the center of the earth.

The energy of waterpower comes from interaction between solar energy and the earth's gravitational forces. This means that waterpower is an energy source that will be available as long as the sun shines and the rains fall.

The whole earth's waterpower resources, if fully developed, could produce five or six times as much energy as was made available by all of the world's electrical generators in 1970. In the United States less than one-fourth of the available waterpower has been developed; worldwide, scarcely more than one-twentieth. Africa and South America, two continents that await industrial development, have nearly half of the whole earth's potential waterpower resources.

Before a country becomes heavily populated, dams across relatively small streams will impound enough water to meet the waterpower needs of the local residents. If the dam and its millpond are established early in the life of the community, the total community environment may be improved by the project. It is another matter to move into an already developed area and build a new dam across a large river with the millpond flooding large sections of the community—sections of the community that provide locations for homes and schools, stores and churches, cemeteries and orchards, farms and gardens, factory sites and public parks.

An untamed river crashing down the cataracts and rocky gorges of scenic mountain country can be a valuable recreational resource. Or it can be the location of a hydroelectric station. With proper planning it could be both. This kind of planning must face situations in which society's energy needs are pitted against other kinds of needs. Supposedly, if the earth's resources are fully developed and adequately managed, the recreational value of its rivers and waterfalls need not suffer from the installation of power-generating stations.

Energy from the Oceans' Tides

The earth's tidal systems could produce the 12,000 trillion kilowatt hours of electrical power needed to drive the whole world's industrial machinery. This assumes that engineers would be able to harvest the tides' energy with total efficiency. This can never be, but there is enough energy in the ebb and flow of the earth's oceans to arouse the urge to tackle the problems that must be solved before tidal energy can be extensively harvested.

Like all natural phenomena that involve gravitational forces, the coming and going of ocean tides are hauntingly mysterious. It takes a good imagination to think of the earth's oceans as enormous bowls of water sticking to a global sphere. The bottoms of these ocean-size containers are as close as possible to the center of the sphere. This means, of course, that when viewed from a perspective outside the earth's surface, some of these bowls appear tilted or even upside down.

Anyone who has carried water in a bowl knows what happens when the bowl is tilted sharply or turned upside down. When water is spilled here on the earth's surface, the water moves as close as possible to the earth's center. The water in the earth's ocean-size bowls also stays as close as possible to the earth's center, that is, the behavior of the water in the earth's oceans is a matter of the earth-centered gravitational forces.

Gravitational forces are universal; and apparently eternal. Even before Isaac Newton stated the law of universal gravitation and ever since, all objects in the universe have attracted each other with forces directly proportional to the product of their masses and inversely proportional to the square of the distance between them. Gravitational forces are not the only forces acting on the water in the earth's oceans. The gravitational forces of the sun, the moon, the stars, and all other objects in the universe threaten to spill the water from the oceans.

The phrase "inversely proportional to the square of the distance between them," however, eliminates much of this threat. Only the moon and sun are close enough for their gravitational forces to compete effectively with those of the earth. And although the moon is many times less massive than the sun, it is so much closer to the earth that its gravitational forces are primarily responsible for the ocean tides.

Each day the rotating earth moves each of its oceans 8,000 miles closer to or farther from the moon. Variation in the moon's gravitational attraction for the water in the oceans in competition with the earth's uniform gravitational forces causes the water in the oceans to be spilled slightly toward or away from the moon. When the rotating earth brings an ocean closer to the moon, the ocean bulges slightly in the direction of the moon.

Coastal irregularities sometimes allow the incoming tide to impound water that can escape only through a narrow channel when the tide goes out. Situations such as this enable engineers to harness the energy of tides. Dams or barriers can be built across the narrow channel. The barriers are opened during incoming tides and closed when a pool of water has been trapped. The trapped water can then be used to drive turbines or other types of waterwheels on its way back to the ocean at low tide.

One such situation is the Bay of Fundy in Maine, Nova Scotia, and New Brunswick. Irregularities in the

Atlantic coastline create more than 500 square miles of tidal pools where water can be impounded. Differences between high and low tides run as high as 30 feet. Engineers say that there is enough tidal energy there to produce more than 5 billion kilowatts—enough to provide electricity for 600,000 people. There are equally promising sites for tidal power stations on the coast of France. And smaller tidal pools, which engineers believe can be used for power production, occur in other countries, particularly England and Argentina.

An American physicist, Albert A. Michelson, has been especially interested in tidal phenomena. He built an instrument that allowed him to study the forces which produce tides. His instrument consisted of a carefully leveled pipe more than 150 feet long which was half-filled with water. Glass windows at each end of the pipe allowed him to observe any changes that might occur in the water level in response to changing gravitational forces.

Michelson's micro-ocean produced tides very much like those in the oceans. He observed that the net effects of the moon's gravitational forces, as well as those of the sun, varied according to the relative positions of the earth, moon, and sun. Because he knew the mass of the earth, moon, and sun and the exact distances separating each of these objects, Michelson could calculate in advance how much his microtides should rise and fall. He found that they rose and fell only 69 percent as much as predicted. This agreed with the difference between calculated and observed ocean tides.

Scientists have concluded from these facts that the whole earth's surface—with all its cities, hills and mountains, highways and forests—rises and falls periodically. It is not only the oceans that ebb and flow under tidal forces. Any water or land area under the moon is pulled up only to sink back down as the moon approaches the horizon. And if the bowl itself is pulled out of shape, this accounts for the less-than-expected spilling of the water in the bowl.

Tremendous amounts of energy are involved in these earth and ocean tides—tremendous amounts of energy that may invite more and more attention when mankind becomes increasingly energy-hungry.

Geothermal Energy Resources

The heat stored in the rocks deep beneath the earth's surface is also attracting increased attention. To some energy experts there is a special challenge in developing ways to harvest the energy responsible for 1000°C. temperatures a few thousand feet beneath the earth's surface in some parts of the world.

Geothermal energy is a particularly inviting resource where local geological conditions permit water to trickle down to the area of hot rocks. In these areas the underground rocks may be porous enough to store dry steam very much the way other kinds of porous rocks store gas or oil. The steam can be harvested merely by drilling wells similar to gas or oil wells. After the bits of rock or sand that come up with the steam have been removed, it can be used to drive the turbines that drive electrical generators.

Recoverable deposits of dry steam do not occur in very many parts of the world. More often the surface water that trickles down into the layers of high-temperature rocks becomes superheated water rather than steam. The temperature is hot enough to change it to steam, but the high pressure caused by the thick layers of overlying rocks does not allow the liquid water to change to steam. Wells drilled in these areas produce water at temperatures as high as 700°F., nearly 500° hotter than the water in a boiling teakettle. Obviously this superheated water will change to steam as soon as it is released from the high pressures created by the rock layers in which it was stored.

The hot springs and geysers of Yellowstone National Park are the best-known geothermal areas in the United States. Dry-steam or hot-water fields have been located in eastern California, Oregon, and Nevada. One 7,000-foot well drilled in California's Imperial Valley produces 700°F. water.

Geologists believe that other geothermal fields await discovery in Alaska, Hawaii, the Gulf Coast, the Appalachian area, and the Ozark and Oachita Mountains. Geothermal energy resources are being developed in many countries, including New Zealand, Japan, Italy, Iceland, Mexico, and Russia. In 1970 the total capacity of the world's geothermal power stations was more than 800,000,000 watts. New construction promised to increase this by more than half during the early 1970's.

171

As is true of so many of the issues involved in the energy shortage, experts differ on the wisdom of investing money to develop the earth's geothermal energy resources. Some critics point out that hot springs and geysers create unique landscapes, which would be seriously damaged by the installation of power stations and the necessary transmission lines.

But all kinds of power stations threaten to damage the landscape. The proponents of geothermal power development add that the hot water, after its heat energy has been harvested, can be used for irrigation and thereby improve rather than damage the surrounding landscape. In those situations where the hot water contains large concentrations of dissolved salts which cannot be removed economically, the spent water can be pumped back down the wells where it came from, thereby avoiding environmental pollution.

The Role of the Earth's "Minor" Energy Resources

The role of wind power, waterpower, tides, and geothermal energy seems to be overshadowed by the more massive roles played by the fossil and nuclear fuels in the dramatic search for energy resources. This has not been always so. And the future may bring a different role.

In many ways these "minor" energy resources show how man's interests in the earth's natural resources change from generation to generation and how these interests influence and are influenced by our overall way of life. A farmer with his own windmill or a village miller with his own waterwheel enjoys a kind of independence that is rapidly disappearing from our highly industrialized society. But the farmer and miller are tied to a provincial way of life. Their do-it-yourself solutions to the energy supply problem would not allow them to expand production enough to compete with industrialized markets.

Traditionally, especially in America, business and industry pay little attention to whether one process or product is more expensive than another from an energy consumption point of view. Nor has there been much thought given to how the availability or scarcity of energy influences the way people live. It has always been a part of the American scene to take for granted the seemingly unlimited energy resources provided on the North American continent.

If people in America live the good life, or if people in some other time or place are forced to live poorly, different forms of government or economic management are most often given as the causes or explanations. But any society is utterly dependent on adequate sources of available energy. If the day comes when the whole earth's energy resources are no longer to be taken for granted —if the day comes when the energy doled out to each person or enterprise must be controlled or limited—then the "minor" energy resources may win increasingly important roles in the drama of human civilization.

8 Getting It All Together

We know some things about the earth's energy supply and what it means to us. At the same time there are many unanswered questions about what we should be doing to head off the problems that will surely develop if our supply of available energy runs short—problems that threaten our hopes and ambitions, in fact, our very survival.

More and More

We know that people, especially in the United States, consume more energy each year than they did the year before. Not only are there more people each year, but each one seems to demand a larger share of the earth's energy.

Some of our society's leaders think that steps should be taken to shut off the increase of our population. This would certainly diminish the demands on our energy resources, and the problem of maintaining an adequate future supply of energy would be at least partially solved. But there is the very big question of how and where population controls are to be introduced and enforced.

Others who are wrestling with the complex problems of life today are asking whether people should expect to have more and more energy at their disposal. They seem to be saying that perhaps it isn't necessary to be constantly seeking to increase the gross national product with the resultant increased strain on our energy resources. The suggestion here is that we will no longer be allowed to use as much energy as we can buy any way we choose. Nor will we be urged to buy as much as we can afford. From now on the earth's supply of available energy will be closely guarded or managed, and everyone will be encouraged to get along with as little energy as possible.

Under this kind of system energy would be distributed or doled out to meet everybody's needs. It is assumed that everyone would realize there is not an infinite supply of energy and would voluntarily let the stream of civilization slow down, that we would all adjust to a lifestyle that doesn't call for more and more energy. There is another highly questionable assumption: Our society would be able to choose among all the things we might want to use energy for and select the uses that would be best for everybody.

Urges to consume more and more energy run through much of what keeps our business and industrial world running—and the governments it supports. Efforts to tamper with these urges could throw the whole system out of gear. There are people who say that if this has to be done, it should be left to the experts. But we are all a part of the whole earth toy. Each of us is vitally interested in the earth's store of available energy. It isn't easy to turn our interests over to the experts no matter how great our confidence in their abilities or how great our faith in their desire to do what is best for us.

People are becoming increasingly concerned about the earth's dwindling supplies of energy resources, especially the fossil fuels. More people are asking more questions about how soon nuclear energy or some other new energy source can be ready to take over. This is good, especially if these people follow up their questions with efforts to gain as much information as is available when their questions can't be fully answered.

Questions of Cost and Convenience

In 1850, 90 percent of the energy used to keep the United States going was obtained by burning wood cut from the nation's forests. By 1900 coal had displaced all but 20 percent of the wood-burning energy business, and oil and gas had taken over about 5 percent of the energy market. By 1950 coal had lost all but about 40 percent of the energy market, and oil and gas had taken over 53 percent. By 1970, 70 percent of the nation's energy was supplied by gas and oil, and only 25 percent by coal. Waterpower held about 4 percent of the energy business throughout these years.

These data suggest that our society, especially its business and industry, can adjust to drastic changes in its sources of energy. Apparently, when a cheaper or more convenient source of energy becomes available, the required conversions can be made. This is reassuring. It is also worrisome. It is reassuring to know that energy has been provided as abundantly, as conveniently, and as cheaply as possible. But it is easy to worry about how long the whole earth's supply of energy will continue to make this abundance, convenience, and low cost possible.

In 1967 the Office of the President of the United States asked a nonprofit organization, Resources for the Future, to study public policies regarding U.S. energy resources. President Lyndon B. Johnson based the request on the need to understand better the nation's future energy needs and the resources that will be available to meet them. The staff members of Resources for the Future brought together the information they believed would be of interest and value to people in industry, universities, national and state government, and foreign governments and international organizations. Their report, *U.S. Energy Policies, An Agenda for Research*, was published by the Johns Hopkins Press in 1968.

Using the language most easily understood by the people who work in government, the report describes how the nation's energy-producing and -selling industries have become established. Included are detailed discussions of how the government has tried to assist or control their efforts to make energy available to the nation. Many of these efforts were prompted by conflicts of interest between one part of the energy business and another or between the public and a private organization or individual.

Especially interesting is the explanation of how the nation's courts settled the question of who "owns" the nation's natural resources. In most countries the minerals and other natural resources underlying the earth's surface belong to the government or are in the public domain. They don't belong to whoever owns the surface land. In countries where the government follows the pattern set by England, however, whoever owns the surface land also owns everything found under the surface.

Because coal stays put, no problems arose from saying that whoever owns the land above coal beds owned the coal. When a coal company bought a piece of land, everyone assumed it had the right to harvest and sell whatever coal it could find and bring to the surface. But oil and gas are fluid and will flow toward any well drilled into their storage areas. In practice one man could harvest the oil or gas that originally existed under his neighbors' property—oil or gas that legally belonged to his neighbors. It didn't take long for the neighbors in such a situation to go to court to seek justice.

As interpreted in the report of Resources for the Future, the courts solved the problem by going back in history and recalling the "law of capture." This law was created to solve the problems that arose when one man killed a deer or other wild game that had its home on another man's property. The argument was that since wild animals were free to roam wherever they choose, they "belong" to whoever captures them. The courts

179

extended this argument and said that oil and gas belong to whoever "captures" them. What oil or gas a man could obtain by drilling wells on his own property belonged to him. It did not matter that the oil or gas might be flowing from under another man's property.

This ruling by the courts caused all the property owners around a newly discovered oil pool or gas field to have wells drilled as rapidly as possible to "capture" their share of the oil or gas. A well on one man's property called for an offset well on the properties of all the neighbors. When an oil pool or gas field was discovered under a town or village with closely spaced building lots, and before the courts came up with a more efficient ruling, a forest of drilling rigs would spring up.

Far more wells were drilled than were needed to harvest the oil or gas efficiently. Often more oil or gas was brought to the surface than could be sold or stored. The excess escaped and fouled streams or created fire hazards. In time the courts came up with new rulings that established safer and more efficient operations. Laws established the minimum size of a tract that could support a well, and all the neighbors shared in the oil or gas harvested from the well. Other laws controlled the amount of oil or gas that could be taken from any well, and special arrangements were made to prohibit the uncontrolled escape of gas from newly opened wells.

Although many oil companies, especially the large ones, were quick to buy up the land where oil pools or gas fields were newly discovered, the general practice was for the landowners to lease their land to an oil- or gas-producing company. Under this arrangement the landowners received a share, usually one-eighth, of the money from the sale of the oil or gas.

The "law of capture" is only a small part of the story of the government's influence, through its lawmaking agencies and courts, on the harvesting of the earth's energy resources. It is reassuring to know that the government can play a role in maintaining adequate sources of available energy. There are also causes for worry. In the case of the "law of capture," the courts turned to the distant past to find a solution to an immediate problem. It did not matter that there were glaringly different conditions involved. It would seem that anyone could have seen the differences between the deer herds that roam a forest and the oil pools that underlie it. For one, the deer herd reproduces itself; the oil pool doesn't. This one difference should have let the courts foresee why deer herds continue to roam the nation's forests whereas oil pools have become extinct in many parts of the country.

How much energy we have used and how we obtained it must be kept in mind when we anticipate the problems of an energy shortage. Similarly, the courts can be guided by former decisions involving the ownership and development of the earth's energy resources. People in general feel more secure when solutions to current problems are based on established practice.

But Americans have assumed in the past that the earth will yield infinite quantities of cheap and convenient energy. This is no longer a safe assumption. Decisions that were good in the past may not be so good in the future.

The future may call for sharp changes in how we look upon the earth's energy resources and how they are to be developed and managed. The days of turning to Mother Nature's storehouses are numbered. And the number of these days is fewer than we would like to believe will be the total span of human civilization.

Faced with the problem of providing adequate quantities of energy for an infinite future, government policies and court decisions may have to swing into new and strange directions. But it will still be up to each of us to do our bit toward keeping these policies and decisions headed in the right direction.

Challenges

Efforts to solve the problems of maintaining adequate supplies of energy throughout the world provide a wide range of challenges. One man may turn his mind and hands to coming up with a totally new energy-producing and -distributing system. Another man may devote his capabilities to finding ways to make better use of the energy supply already available. Or to create a new lifestyle that would allow us to live just as well but use only as much energy as will be continuously available.

J. O'M. Bockris, an Australian physicist, provides a good example of one type of response to the challenge of an impending energy shortage. He describes in *Science* for June 23, 1972, how hydrogen could be used to replace petroleum fuels in nearly all their present-day uses. He foresees hydrogen used to carry energy from its production in nuclear reactors to wherever it is needed, very much as today's copper and aluminum cables and wires carry electrical energy from power stations to customers.

Bockris' plan calls for nuclear reactors located on platforms floating on deep water. This location avoids thermal pollution of the environment. The energy produced by the reactors is used to separate water by electrolysis into its hydrogen and oxygen components. The hydrogen is to be piped or carried in tanks to distribution centers and thence to all kinds of industrial and domestic, wholesale and retail customers. Each customer, in turn, will have fuel cells that use the hydrogen and, supposedly, oxygen from the air to generate electricity with pure water as the only byproduct.

Automobiles, trucks, trains, and ships will use power systems consisting of fuel cells to charge batteries that drive electric motors. These systems are to be nonpolluting, silent, and much more efficient than today's petroleum-fueled internal-combustion engines. Airplanes will be fueled with hydrogen jet engines, and these lightweight, highly efficient engines will allow airplanes to travel much farther on the same weight of fuel. In further contrast to petroleum-fueled jet engines, pollution of the atmosphere with carbon dioxide, nitrogen oxides, and other smog-forming gases will be avoided.

Bockris lists three things that will delay putting his plan into practice: (1) conservatism; (2) the absence of education or training in electrochemical engineering; and (3) the public's fear of hydrogen. Problems such as these ought to be solved more easily than the problems that would flare up if the nation's gasoline pumps ran dry or its electrical power stations were browned out.

In 1970, nearly 70 percent of the plastics, synthetic rubber, textile fibers, detergents, medicines, dyes, explosives, and other products of industrial organic chemistry used petroleum as the source of raw materials. The petrochemical industries used 250,000,000 barrels of feedstock. To show further how the depletion of the nation's petroleum resources would create serious problems, in 1970, the nation's refineries provided more than 148,000,000 barrels of the asphalt used to build streets and highways.

All of the Earth's Resources Are Finite

Bruce M. Hannon's response to the challenges associated with the threat of an energy shortage illustrates the interdependence of all of the earth's resources. In the March, 1972, *Environment* Hannon compares the amount of energy that must be spent to use throwaway rather than returnable beverage containers. He touches on the differences in the amounts of steel, aluminum, and glass that are consumed in making both kinds of containers. The earth's rocks contain finite quantities of copper, silver, aluminum, gold, tin, sulfur, arsenic, phosphorus, and all the other elements from which we build the material world. In most cases more energy is needed to recycle these resources than to take the raw materials from their original storage in the earth's rocks.

It would take another book to tell the whole story of

how we must adjust our ways of living to fit the relative abundance or shortage of the earth's material resources. But the stories of the earth's natural resources, material and energy, are very much interdependent.

Each step in making the various kinds of beverage containers can be "charged" for its individual energy consumption. To mine the iron ore to make the cans to carry one gallon of soft drink, for example, calls for 1,570 BTU's. To make the aluminum lids for these cans calls for 12,040 BTU's. To transport the filled cans from the bottler to the retail store calls for 6,400 BTU's.

When Hannon added up the total energy charge for enough throwaway cans to carry one gallon of soft drink, the bill came to 51,830 BTU's. For comparison, if returnable glass bottles are used (assuming each bottle makes eight round trips before it is destroyed), the total drain on the earth's energy supply comes to 19,970 BTU's per gallon of soft drink delivered to customers.

The energy demands for returnable versus throwaway glass containers are especially revealing. When sixteen-ounce returnable bottles are used to deliver one gallon of soft drink, the total energy demand is 19,970 BTU's. When throwaway glass bottles are used, the total energy demand jumps to 62,035 BTU's. Approximately 4,000 BTU's can be saved, however, if the new throwaway bottles are made from recycled throwaway bottles.

In the 1960's, throwaway containers for soft drinks became a popular convenience. People seemed to pay little attention to the greater demands on the earth's energy and material resources or to the extra cost, including the greater cost of keeping our environment clean. Hannon's investigation seems to point the way toward increasing concern for the earth's dwindling supply of energy and material resources. Similar studies of the energy and material demands of all the gadgets we use and the things we do would put society on its way toward handling many of its critical energy and material resource problems.

There's a Role for Everyone

When faced with the problem of finite supplies of energy against the infinite demands of today's society, different people come up with widely different responses. One man tries to develop a whole new energy-supply and -distribution system. Another turns his mind to making the best possible use of the energy resources still available. One person may try to consider the threat of an energy shortage as it would influence the relationships among nations. Another may pay more attention to his individual energy needs and what a shortage of energy would mean to his preferred life-style.

The crucial point is that people are concerned. If the day comes when there simply isn't enough energy available to keep the whole earth toy running the way we want it to run, we will be ready to make the most of what we have. This is the response of intelligent, enlightened, socially sensitive people who are concerned about what they expect and want to do in today's giant windup toy—and tomorrow's.

186

INDEX

187

The Author

John H. Woodburn has published six other books related to science, including the popular *Know Your Skin*
from Putnam's. Since serving as an air navigator in the
U.S. Navy during World War II, Mr. Woodburn has
been a writer and has taught in both colleges and high
schools. He holds a doctorate degree and he and his wife
live in Chevy Chase, Maryland.